Ethical Adaptation to (

Ethical Adaptation to Climate Change

Human Virtues of the Future

Edited by Allen Thompson and Jeremy Bendik-Keymer

The MIT Press
Cambridge, Massachusetts
London, England

MIT Press books may be purchased at special quantity discounts for business or sales promotional use. For information, please email special_sales@mitpress.mit.edu or write to Special Sales Department, The MIT Press, 55 Hayward Street, Cambridge, MA 02142.

This book was set in Sabon by Toppan Best-set Premedia Limited. Printed and bound in the United States of America.

Library of Congress Cataloging-in-Publication Data

Ethical adaptation to climate change : human virtues of the future / edited by Allen Thompson and Jeremy Bendik-Keymer.
  p. cm.
Includes bibliographical references and index.
ISBN 978-0-262-01753-4 (hardcover : alk. paper) — ISBN 978-0-262-51765-2 (pbk. : alk. paper)
1. Climatic changes. 2. Global environmental change. 3. Human beings—Effect of climate on. 4. Restoration ecology. 5. Environmental ethics. 6. Environmental justice. I. Thompson, Allen, 1969– II. Bendik-Keymer, Jeremy, 1970–
QC903.E83   2012
179'.1—dc23
2011038545

10 9 8 7 6 5 4 3 2 1

For Abraham and Annabelle—AT
For Isaiah—JBK

Contents

Acknowledgments

This book began with a conference, "Human Flourishing and Restoration in the Age of Global Warming," held in September 2008 at Clemson University and sponsored by the College of Architecture, Arts and Humanities, the Restoration Institute, the Science and Technology in Society program, the Sustainable Universities Initiative, the Rutland Institute for Ethics, and the Department of Philosophy and Religion. I thank these university sponsors and all of the conference participants for making that event so successful. I also thank Clay Morgan from the MIT Press for his interest and help and Kathy Caruso and Julia Collins for all their work improving the text. The publishers of *Science, Technology, and Human Values* generously gave us permission to reprint Dale Jamieson's "Ethics, Public Policy, and Global Warming."

More personally, I thank my wife, Pamela, my parents, Ken and Joan Thompson, and my friend Jeremy Bendik-Keymer, from whom I continue to learn.

—AT, Corvallis, Oregon, August 2011

I want to thank Elaine Wolf, Esther Bendik, David Keymer, Jim Davidson, all the volume's contributors—especially Breena Holland for co-organizing the conference to which Allen refers—and Allen Thompson for being an example of someone who has lived, for decades now, with an eye to the virtues of the future.

—JBK, Cleveland, Ohio, August 2011

Introduction: Adapting Humanity

Allen Thompson and Jeremy Bendik-Keymer

Mitigate we might, adapt we must.
—W. D. Nordhaus, *Managing the Global Commons*

To adapt requires learning from our mistakes. As the environment changes, so must we.
—Kenneth Shockley, chapter 14, this volume

We acquire the excellences through having first engaged in the activities. . . . We become just by doing just things, moderate by doing moderate things, and courageous by doing courageous things. What happens in cities testifies to this; lawmakers make the citizens good through habituation.
—Aristotle, *Nicomachean Ethics*

The Problem and the General Idea

When we take all of our core values seriously, they can make tough demands on us. This is the case regarding climate change, where our values ought to commit us to a robust understanding of adapting ourselves to a new global climate. As many have argued, climate change is not just a scientific or economic matter but is an *ethical* one as well (e.g., Gardiner 2011; Gardiner et al. 2010; Jamieson 1992, revised and reprinted as chapter 9, this volume). Focusing on ethical concerns, we may ask "what *should* become of us under global climate change?" In light of this question, our volume is about a robust and normative understanding of adaptation to anthropogenic global climate change.

Given current predictions about global climate change, one finds increasingly stark scenarios of environmental catastrophe (Lovelock 2006). There are also attempts to buffer our contemporary form of life from the ill effects of climate change so that we can more or less continue with business as usual. Against and in between both trends, suppose we thought about how we might *flourish* in a new global climate? Suppose, too, that we understood flourishing as the ancients did. Suppose it were thoroughly ethical, moving beyond conceptions of subjective well-being to involve a broad conception of human excellence, beginning with the capabilities needed to live a life with dignity and moving onto virtues and vices of a new world, even well-worked-out relationships between our lives, our institutions, and the extrahuman world. Flourishing would involve a just and attentive relationship to our environment, rather than an unintentionally destructive one, and would involve *restoring* aspects of the environment we had unintentionally destroyed or accepting the novel ecosystems we have unintentionally created (Hobbs 2006). To flourish in a new global climate would then include more than adapting to radically new environmental conditions *in order to carry on* much as we had before. It would eschew desperate and paralyzed resignation to impending catastrophe (Hamilton 2010). It would mean changing our sense of ourselves, especially our virtues and vices, and finding what kinds of social conditions best clarify and support our dignity and decency.

There is no shortage of how-to guides emerging in response to climate change (Steffen 2006). These give readers concrete tips on how they can change their behavior to make it more sustainable. There is also a growing movement of environmental virtue ethics. It examines how our characters can become ecologically attuned. Then, in subfields as diverse as ecological restoration, human capability studies, organizational theory, and adaptive management, scholars currently explore how diverse aspects of our lives in and outside of institutions, in urban or wild contexts, might handle climate change responsibly, attentive to a variety of natural values. These disparate subfields have much to learn from each other. They seem to work on different facets of the same project: adapting ourselves to the realities of anthropogenic global climate change.

Yet even the most innovative how-to guides and these disparate subfields suffer from a fragmentary view of the whole. Underlying assumptions about our natural context, character, institutional limits, or organizational possibilities remain out of sight to each of them in different ways. Our volume addresses this shortcoming by bringing together research in important subfields of environmental philosophy and political theory with an eye toward adapting our human flourishing, and especially our character, to the expected realities of a new global climate. It does so through a set of focused and cutting-edge debates about the role of history in ecological restoration under climate change, the relation between (human development) capability theory and the environment, the precise nature of environmental virtues under climate change, and the tangled and often contradictory status of institutions in enabling or limiting human flourishing and specifically virtue. Through evolving, back-and-forth discussion, it explores how our understanding of human flourishing ought to change as a result of climate change.

Our purpose in this introduction is four-fold. First, we want to distinguish and explain how our view of adaptation to climate change is committed to reflection and research on human flourishing. Second, we want to explain how our approach to virtue theory and the subfields we've assembled here contribute to thinking about adaptation. Since both our understanding of virtue theory and our understanding of adaptation are not standard to the discourses surrounding them, we take pains to mark out how this collection contributes novel work to each area. Third, we want to explain how thinking about the role of virtue in ecological restoration connects with adaptation and human flourishing under global climate change. Finally, we want to articulate how the sections and chapters in this collection contribute to an overall argument, thus laying out the challenge of adapting humanity.

Thematic Overview: Adaptation, Excellence, and Restoration

Adaptation

We begin by distinguishing our view of adaptation. It is well known that various human activities, including deforestation and burning fossil fuels for energy, are the fundamental cause of both current and significant future changes in global climate conditions. Good accounts of the scientific basis for this claim, and descriptions of the consequences that climate change may have for human and biological systems, are widely available and will not be reviewed here. Faced with the specter of global crisis, in 1992 the UN enacted the United Nations Framework Convention on Climate Change (UNFCCC) was established in 1992 in order to facilitate coordination of an international response to anthropogenic climate change. The text of the Convention established two main categories of response: mitigation and adaptation.

According to the Intergovernmental Panel on Climate Change: "Mitigation (or "limitation") "attempts to deal with the causes of climate change. It achieves this action through actions that prevent or retard the increase of atmospheric greenhouse gas concentration by limiting current and future emissions from sources of greenhouse gases and enhancing potential sinks" (IPCC 1996, 831). Mitigation aims at the "stabilization of greenhouse gas concentrations in the atmosphere at a level that would prevent dangerous anthropogenic interference with the climate system" (UNFCCC Article 2). The purpose of mitigation is to prevent further damage.

To date, the international efforts at climate change mitigation are inadequate, and consequently we are beginning to see serious attention turned to issues of *adaptation*. Despite the establishment of legally binding emission limits in the 1997 Kyoto Protocol, contemporary global greenhouse gas emissions are at the very high end of projections considered in the 1990 International Panel on Climate Change First Assessment Report (Moore 2008).[1] Global climate change is occurring now, damaging natural systems and harming human beings, and we are already committed to more warming. At present, there is no agreement on whether and how to extend the Kyoto Protocol, beyond the end of its first commitment period in 2012, nor is there a new treaty ready in the wings to replace it. As efforts to mitigate climate change fail to prevent climate damage, the 2010 UNFCCC meetings in Cancun saw agreement on an adaptation action plan intended to prepare people for living with the environmental and social challenges caused by global climate change.[2]

According to the IPCC, adaptation is concerned with responses to both the adverse and positive effects of climate change. It refers to any adjustment—whether passive, reactive, or anticipatory—that can respond to anticipated or actual consequences associated with climate change (IPCC 1996, 831). With this increased attention, the concept of adaptation has become a source of significant debate in academic and policy circles, concerning both its meaning and practical interpretation (Schipper and Burton 2009, 2).[3]

The concept of adaptation involved in mainstream policy discussions was prominent at both the UNFCCC 2009 meetings in Copenhagen and the 2010 meetings in Cancun. Representatives of those nations most vulnerable to harms caused by climate change articulated demands for the developed countries to provide financial assistance and clean-energy technologies to vulnerable, less developed nations.[4] According to the appeal, these aids could help nations with some of the world's poorest people to *adapt their economies* to damaged ecosystems and new emission restrictions. Indeed, it seems prudent to prepare for the range of bad consequences from global warming that we are already committed to (and perhaps worse potential consequences). In addition, there are strong grounds of justice to support these goals (Shue 1999).

As the concept of adaptation appears in international climate policy negotiations, there is no doubt that it raises important normative questions. For example, relevant issues under discussion at the UNFCCC's 16th Conference of the Parties in Cancun concerned the legitimacy of various claims for adaptation funding,[5] the establishment of an appropriate institution to oversee and administer adaptation funds, and, of course, the source of these funds themselves.[6]

Yet this concept of adaptation seems incomplete, even misleading. Is adaptation simply a matter of technology and funding? The kind of adaptation currently rising to greater prominence in international policy negotiations is distinct from—and less difficult than—another type of adaptive challenge that anthropogenic climate change presents to us. Funding and technology transfer from nations with developed economies to vulnerable nations with only developing economies are designed (wherever or however those resources are targeted) to enable poor nations to continue along a path of economic development. That path is traditionally conceived to advance human well-being by increasing the nation's economic productivity.[7]

Adaptations here are understood as enabling economically developing countries to retain an upward trajectory despite climate change's effects. In this sense, adaptation is really an effort to *mitigate* the impacts of global warming. It is trying to buffer economic development and the people dependent on it from extreme consequences. Adaptation thus becomes another name for mitigating the impacts of what we have brought about on the form of life that we presently lead. Just so, adaptation has been defined by the Pew Center on Global Climate Change as "Actions by individuals or systems to avoid, withstand, or take advantage of current and projected climate changes and impacts. Adaptation decreases a system's vulnerability, or increases its resilience to impacts" (Pew Center 2007, 3).

The Pew Center's definition might be summarized as the view that *adaptation only reduces vulnerability*.[8] According to Schipper and Burton, formulations such as these are common in the literature, but are a detrimental inversion of Kelly and Adger's point that "adaptation *is facilitated by* reducing vulnerability" (Schipper and Burton 2009, 4; see also Kelly and Adger 2009, our emphasis). But should adaptation be conceived of as *only* reducing vulnerability?

We believe this conception of adaptation, dominant in policy negotiations, is too narrow.[9] From our perspective, most contemporary discussions of adaptation are closer to coping strategies.[10] They are merely short-term responses; they are not sustainable (Davies 2009). Accordingly, we believe that the environmental circumstances that exist today pose a different and greater challenge to human adaptation. We call our view a *humanist view of adaptation*. It can be seen by contrasting two views of adaptation that Burton identified as the "pollutionist" and "development" views (Burton 2009, 89).

Both the pollutionist and development views are species of what the Pew Center advocates—namely, that adaptation only reduces vulnerability. However, in each case, the focus is different. The pollutionist view focuses on reducing our vulnerability to climate change *in a way that allows* higher pollution levels than would otherwise be acceptable. According to it, adaptation enters the UNFCCC's objectives "because the more adaptation can be used to reduce impacts that might be considered dangerous, the higher the threshold of [greenhouse gas] concentrations that can be accepted" (Burton 2009, 90–91). This perspective may appear cynical, but it has the merit of being realistic about the capacity of the global economy to substantially wean itself of fossil fuel dependence. While few would openly acknowledge holding a pollutionist view, it's also clear that many interested and well-represented parties strongly resist any serious efforts to decarbonize the global economy while still acknowledging that something should be done to assist those most vulnerable to climate harms.

The development view of adaptation, by contrast, is more widely held. Here adaptation is pursued *for the sake of* buffering human development, which is at risk under conditions of global climate change. Accordingly, adaptation must be incorporated into development planning and implementation in an uneasy tension with the projected affects these changes may have on the acceptability of different greenhouse gas concentrations. On this view, most adaptation just is a form of development, because protecting vulnerable populations involves improving their health and educational systems, for example, which in turns builds the community's resilience to both direct and indirect climate harms. However, projects that would not proceed in the absence of climate change, such as building sea walls, are understood as advancing development under new conditions. The development view is thus more focused on protecting human capabilities than preserving business as usual. Yet even it conceives of adaptation only as a means to reduce human vulnerability to climate change, the imperatives of development remain unchanged.

We believe both the pollutionist and development views are short-sighted, largely because they do not examine the understanding of human flourishing that is implicit in their respective notions of economic and human development, which serve either as givens or as benchmarks in their views. We think that who we are as humans and how we understand flourishing need adjustment in order to for us to adapt *well* to a new global climate and to be responsible people. Accordingly, in place of these two views, we advocate:

The humanist view of adaptation   Instead of attempting to mitigate the impacts of climate change on who we are at present, adaptation means adjusting our conception of who we are to appropriately fit the new global context.

Contemporary and projected global conditions present us with the challenge neither only of preparing ourselves to avoid or withstand climate-related changes, nor merely of aiming to protect our contemporary form of life (which many authors in this volume explicitly claim to be unsustainable), but rather of adjusting our sense of ourselves, our ideals, our understanding of responsibility, relevant virtues and vices, and the scale and structure of our institutions—to take into account the realities and exigencies of the new world in which we find ourselves. This is the challenge of adapting humanity itself. The challenge so stated is, of course, quite broad:

Adapting humanity itself   Adjusting (1) who we are and (2) how we enable ourselves to (3) best meet new and projected ecological conditions, including climate change and instability. In the face of these, understanding human flourishing in new ways.

Three concepts stand out, that of (1) who we are; that of (2) enabling ourselves; and that of (3) what is best in our response to climate change. *Who we are* is a question of our moral character and our social institutions. It is a question that is central to parts III and IV of this volume, "Adjusting Character to a Changing Environment" and "Reorganizing Institutions to Enable Human Virtue." How we *enable ourselves* is a question of capability, which is a central focus of part II, "Integrating Ecology into the Virtue of Justice," and it is a question also addressed by part IV. The question of how we can *best* meet new and projected ecological conditions is both a pragmatic matter that restoration theorists discuss at length in part I, "Adapting Restoration to Climate Change," and a question of value that reappears throughout parts I–IV. Together, the four parts, focusing on who we are, how to enable ourselves, and how best to meet climate change, contribute to understanding human flourishing in new ways. Instead of only adapting our infrastructure and economies to mitigate damage, we believe the challenge of adapting to global realities today and in the future will require humanity to revisit and suitably adjust the practice of ecological restoration; the place of ecology in our conception of justice; the form and substance of some traditional virtues and vices; and the organization, scale, and underlying metaphors of important institutions. Accordingly, these topics form the organizational backbone of this book.

The sixteen chapters here explore the idea that the challenge of adapting to global climate change is fundamentally an ethical challenge of adjusting our conception of humanity, that is, of *understanding human flourishing in new ways.*

Excellence

We have explained how the humanist view of adaptation depends on reflection and research on human flourishing. But our initial account is still quite schematic. What, more specifically, does human flourishing involve?

One way to get a grip on human flourishing—which does not reduce it to subjective well-being but instead reflects a life worthy of human dignity (Holland, chapter 7, this volume)—is through the ancient philosophical notion of *excellence.* On the humanist view we offer, adaptation to global climate change is open to forms of ethical discourse that are distinct alternatives to ethical analyses concerned primarily with utilitarian calculation, international obligations, and distributional equity. In particular, the evaluative concept of excellence comes to the fore. We believe the humanist view of adaptation involves the notion of excellence in two distinct but interconnected ways.

First, the *processes* of adapting ourselves—individually and collectively—can be done well, or not. Human beings as a biological species are flexible and widely adaptive. Thus under conditions that call for it, the process of adjusting ourselves to fit new conditions can itself be a form of human goodness. Humans who adapt exhibit a kind of natural excellence, while the failure to adapt exhibits a form of deficiency or malfunction. The human good of actually adapting (or perhaps, the good of adaptability) is a standard against which we can evaluate human beings, both individually and collectively. We believe this kind of excellence in adaptation is important, but it is dependent on the second way excellence can figure in specifically human adaptation, which determines whether "successful adaptation" is simply survival or is, instead, suitable as flourishing.

The adaptations that good human beings will undergo can themselves be more or less suitable. This is the second way excellence figures in adaptation. Instead of being obstinate, it's good that we adapt. But furthermore, while different adaptations are possible, not all adaptations are equally befitting or appropriate. Under a set of changing circumstances, human excellence appears when we transform ourselves in ways that we can reasonably judge are well suited both to ourselves and to the new circumstances. There is an ethical notion here of what is an *appropriate* adaptation, of something that is *well suited*. Here, the question is left open of whether protecting the status quo could be enough.

Whereas the adaptive responses of other biological systems to environmental change are entirely reactive, it is important to recognize that "the responses of human systems are both reactive and proactive. . . . Human systems may adjust in pursuit of goals other than mere species survival, for example, to enhance quality of life" (Smithers and Smit [1997] 2009, 19). As mere animals, good human beings will adapt reactively. But as specifically human animals, we can take the opportunity to calibrate our environmental character, practices, and institutions, because our *values* demand it, because we could not live with ourselves otherwise.

A proactive, value-driven form of adaptation may, in more ordinary circumstances, require only relatively superficial changes to our patterns of behavior. But the prospect of radical and irreversible anthropogenic global climate change suggests we are talking about extraordinary circumstances, ones in which a value-driven adaptation may ask us to pursue the very revaluation of our values, including our moral self-conception as humanity. If we do so, we adapt by reconceiving our humanity, with an eye to the possibility of living well under the radically new environmental conditions that we are responsible for. New forms of excellence come to the fore as being helpful or even crucial to a human life worthy of dignity. This is to understand our flourishing in new ways. The chapters in this volume concern this second sense of adaptive excellence. They begin to outline more precisely how we could proactively adapt our self-conception, and begin adjusting ourselves to live life well through the emerging Anthropocene Epoch (Crutzen 2002)—a name proposed for the geological epoch in which Earth ecology substantially shifts as a result of *Homo sapiens*.

So concerned with human excellence, the chapters in this volume fit comfortably within the tradition of virtue ethics based on the concept of *eudaimonia*, or the well-lived human life. In this tradition, virtues are excellences, and excellence in living well puts one on the trail of candidate virtues. "Eudaimonia" is often translated as "human flourishing"— an expression used frequently in this volume. Each chapter concerns, according to different dimensions, human excellence in adapting ourselves to the challenges posed by global climate change. In this way, the collection presents a contribution to the growing body of literature in environmental philosophy known as *environmental virtue ethics* (Sandler and Cafaro 2005; Sandler 2007). The English word "virtue" has an origin in the Latin "virtus," as moral perfection, which was a Roman translation of the Ancient Greek term for excellence, *arête*. Thus, the moral virtues are the excellent, or good-making, character traits of a human being qua human, and *eudaimonia* (human flourishing or the well-lived human life) is the name for a life lived well on the basis of these excellences.

A broad conception of human excellence, including traits of character and the well-lived life, is a primary ethical lens through which chapters of this collection examine the challenge of adapting humanity. In part I, "Adapting Restoration to Climate Change," authors consider the place of character virtues, both old and new, in the practice and goals of ecological restoration. Broadening, or adapting, the virtue of justice for new global environmental conditions is the central topic of part II, "Integrating Ecology into the Virtue of Justice." And perhaps most straightforwardly, character virtues and vices relevant to climate change are the focus of part III, "Adjusting Character to a Changing Environment." But the story does not end there. With part IV, this volume makes perhaps the most original contribution to the literature on environmental virtue by shifting the focus away from a modern, narrower sense of virtue to a more ancient, and broader, sense, whereby institutional organization and thought about our enabling conditions are part and parcel of adequately grasping human excellence.

In effect, we underline how character virtue is one *dimension* of human flourishing, a part that must be conceived in conjunction with broader, more collective, and more ecological aspects of our flourishing. We place character virtue within various contexts—including the practice of ecological restoration, the capabilities of living things and living systems, and the politics and organization of human institutions—as being inseparable from them. And all of these are couched within the ecological condition of our life on Earth.

Doing all this, we believe, is an important corrective to contemporary virtue ethics. Take first the point about institutions. Although virtue ethics takes its inspiration from ancient philosophy, much contemporary virtue ethics is, ironically, constrained by an overly private or individualistic focus that is a result of our modern form of life. Aristotle's *Ethics* (Aristotle [ca. 350 BCE] 2002) was a part of his *Politics* (Aristotle ca. 350 BCE), as indeed his teacher's *Republic* (Plato ca. 380 BCE) was simultaneously the first moral psychology of character in Western philosophy and the first work of political theory. Some modern authors such as Rousseau (1762a, b) and Hegel (1820) sensed or fully grasped the interconnection of the moral and the political, but contemporary environmental virtue ethics has not fully done so yet.

The reason why is that we are often products of our time. Modern shifts in thinking changed the scope of our ideas about the ethical domain, effectively limiting our ideas about ethics to concerns about causing harm to others. In effect, the ethical domain became the moral one, a narrow domain of interpersonal concerns. These shifts in thinking came to a head in the separation of the private from the public sphere, elucidated most clearly by Benjamin Constant's (1819) famous discourse on the liberty of the ancients verses the liberty of the moderns, and especially in the moral philosophy of John Stuart Mill (1859). One consequence of these shifts in the scope and object of virtue is a modern tendency to think of human virtue in moralistic terms, and, of course, to focus on human individuals and their other-regarding traits of character rather than on virtue as embedded in and inseparable from the enabling conditions of a social and political context. Our volume fills the gap created by this overly moralistic conception of virtue and does so with good reason: problems like climate change involve global collective action and bring to the fore the way our individual lives are enmeshed in and constrained by institutions, including the ways we organize our lives and our politics.

Thus, taking our cue from the ancients, we believe it is crucial to contextualize human character within the institutional nature of human life. In part IV especially (although it is a theme that recurs at various points throughout the volume), reflection on the scale and structures of collective action that characterize global environmental problems like climate change leads many of our chapter authors to the realization that scholars interested in character and human flourishing ought to focus on how we are caught up in our institutions. We humans live socially and collectively. Our relations with our ecological environment are largely mediated through human institutions. Who we take ourselves to be rests on assumptions about what is institutionally ordinary, given, or possible. Accordingly, our institutions are a part of our humanity, a part of *where* and *how* we are that also must be adapted well to our changing environmental reality. As Steven Vogel points out in chapter 15, organized society is our essential human environment within Earth's ecology.[11]

But expanding our conception of virtue, as the ancients and some moderns did, to include human social and political institutions is simply not sufficient, we believe, in the face of contemporary environmental challenges. We must take a step beyond and consider also how human virtue is embedded in our relations to the Earth's ecology. No doubt, excellent adaptation ought to involve work on personal virtues and vices, as is currently underway in environmental virtue ethics. But since we want to move beyond the moral sphere to conceive of human flourishing broadly, it is also crucial that we think of human excellence *ecologically*.

This takes us to our second point, about embedding virtue in our ecology. Not only did this escape the moderns but it eluded the ancients as well. One cannot find more than a passing awareness in the works of either Plato or—more surprisingly—Aristotle that a large part of understanding human virtue ought to be grasping how it relates to the enabling conditions of, and wider relations we have with, Earth's ecology (Holland, chapter 7, this volume; Schlosberg, chapter 8, this volume).

But climate change forces such an awareness on us—at the very least by the way the reshaped macroecological conditions bear down on human life, and the way global environmental problems reveal to us our massive capacity to bring unintentional destruction to other forms of life. Accordingly, there is a real and pressing need to render virtue suitably embedded in its ecological context and to work out our relation to the wider sphere of life on this planet. Both imply thinking of human excellence *ecologically*. That is, we must see *both* to what extent our ecological nature shapes the map of human life into which virtue fits *and* to what extent human excellence ought to extend outward to encompass our relation with other forms of life and with the ecologies we affect. Thus the chapters in parts I and II especially (although this theme runs throughout the volume) concern how we can incorporate parts of the Earth's ecology and our relations with them—from restoring landscapes and caring for individual nonhuman animals, through conceiving of ecosystems as communities, to managing the global climate system itself—into our conceptions of human flourishing.

Thus, the work gathered in this collection moves our thinking about the environmental virtues from a conception of individual character excellence to a broader, richer, and more complex conception of the excellence of human individuals who also share a collective life as members of a society, of common humanity, and of the broader community of life on Earth.

In closing, let us review how the chapters collected here hold together in a project of bringing a virtue-theoretic orientation to bear on issues of adapting to global climate change. First, by introducing the humanist view of adaptation, in contrast with a view of adaptation as only reducing vulnerability, we make room for the relevant ethical discourse to move beyond questions of equity and obligation. Second, by contextualizing the ethical ideals of human excellence in terms of ecology, practices, capacities, and social and political institutions, we extend concern with human virtue beyond a focus on character traits and individual flourishing. Thus, our volume does not propose to explore how character virtues such as generosity or benevolence, for example, may offer guidance regarding the source or distribution of adaptation funding for, say, desalinization plants in communities where rising sea levels threaten groundwater supplies. Instead, the theme unifying these chapters is exploring how excellence in human adaptation in the face of an unfolding climate crisis leads us to understand human flourishing in new ways.

Restoration

The philosophical dialectic that characterizes the focused and connected conversations of this volume begins out in the field, where restorationists reflect on human character and come to terms with how climate change may alter the basic conception of ecological restoration. We want to say something about this starting point.

Ecological restoration is "the process of assisting in the recovery of an ecosystem that has been damaged, degraded or destroyed" (Clewell, Aronson, and Winterhalder 2004). It is a practice. It is related to *restoration ecology*, which is a body of knowledge. Restoration ecology is a part of conservation biology concerned with the application of ecological principles in the practice of restoration. Although ecological restoration intimately involves knowledge of restoration ecology, our interest is with the former, that is, with those intentional practices that aim at attempting to rehabilitate ecosystems which previous human activity has damaged, degraded, or destroyed.

Ecological restoration is often traced back to work done at the University of Wisconsin's Curtis Prairie, under the leadership of Aldo Leopold. While ecological restoration began as isolated pioneering experiments, today it is a thriving industry engaged in widely by both the private and public sectors. Restoration projects are manifold and vary in scale, from the recovery of ecosystems in small urban lots to rehabilitating significant portions of an entire bioregion—for example, the state and federal project of restoring the Florida Everglades.

However, the projects and practices of ecological restoration vary in ways other than scale. As any specific restoration project is tied to a particular place—that is, the specific site to be rehabilitated—details particular to different restorative practices will vary accordingly. What kind of ecosystem is to be restored will affect the relevant strategies and practices. So will details about what the disturbance was, where the ecosystem is being restored (including both its natural and social contexts), and perhaps most important *why* it is being restored. In sum, ecological restoration involves a widely diverse set of behavioral practices that may hang together only by family resemblance.[12] Restorative projects are undertaken for different reasons and often follow different methodologies, at least in part because they are undertaken by different people, in different cultures, and in different times and places. But as natural ecosystems themselves exhibit diversity, we should not be surprised to find that efforts aimed at restoring them are likewise diverse.

Ecological restoration is the growing edge of how we actually deal with the undesirable effects of human life on Earth's ecology. Its firsthand involvement is repairing damage we have caused. Through restoration, we learn about the practical limits and possibilities of reincorporating ecosystems in our lives, even if our action is a kind of restorative justice that ends up leaving a revitalized ecosystem alone. Through restoration, we also learn about where we can and can't expect to find success in our efforts to construct, care for, and manage complex ecological systems. We learn the dimensions of our capability to rehabilitate nature. Perhaps more significant, we also learn in the process something of *who* we need to be, in order for our interventions to be successful—by some measure—and sustainable (Throop, chapter 2, this volume). Thus, at the very least, the knowledge we gain through the practice of ecological restoration is helpful for understanding our role in recovering and then managing the integrity and functioning of ecosystems that we have significantly damaged. At the most, it helps us move closer to being a "plain member and citizen" of the Earth's biotic communities (Leopold 1949).

Anthropogenic climate change is causing and (mostly likely) will increasingly cause significant damage to ecosystems worldwide. The realities of climate change will make it necessary for human beings, collectively, to adaptively manage the basic ecological conditions of the global biosphere. This will be done through intentional actions but also through oversight and omission, by vehicles of international policy, by large-scale economic and other social institutions, and by various ordinary but widespread human practices. At present, it appears that we—humans—are not ready to adopt this role and its attendant responsibilities. *Who we are* today is not ready for this and *who we have been* got us into this mess. Since it seems that who we are will have to change, lessons we can learn from ecological restoration about responsible ecosystem rehabilitation and successful management will be invaluable. In particular, what we learn about the effects of global climate change on our understanding of ecological restoration should be transferable, *mutatis mutandis*, to our thinking about stewarding aspects of the global biosphere, and to being responsible for it, under the same conditions.

Climate change affects our conception of human excellence in the practice of restoration. It shows us that *we* have to change. Ecological restoration does this in two ways. First, borrowing on its now well-established hopes for articulating how we could develop a good relationship with nature, ecological restoration shows us the virtues we ought to acquire if we wish to move from a damaging and alienated relationship with our environment to one that can allow both us and the world of life around us flourish. Second, through an emerging debate surrounding the value of historical fidelity in setting restoration goals, it shows us that our relationship with nature, with the world we collectively make and live in, must be reconciled with the realities of climate change.

Overall, it is plausible to believe that among the important and ethically appropriate responses to climate damage should be some that are modeled on the practices of ecological restoration, but scaled appropriately (Hirsch and Norton, this volume). Far from the current rush toward geo-engineering, this kind of response would exhibit the virtue of humility (Throop, chapter 2, this volume; Gardiner, chapter 12, this volume). And we might even conceive of our restorative activities as a demand of justice (Keulartz and Swart, chapter 6, this volume; Schlosberg, chapter 8, this volume; Holland, chapter 7, this volume). The possibility of realizing human excellence in the practice of rehabilitating damaged ecosystems would then illuminate a new form of human flourishing exhibiting an excellent relation to the environment. And this would likely illuminate aspects of adaptive excellence generally.

Argument Overview: History, Justice, Virtue, and Context

Human flourishing is a very broad concept. Accordingly, it is usually best approached through specific contexts. In this volume, we track a number of focused debates that prove insightful for adapting humanity's sense of itself, and which speak to each other in often surprising ways. What develops across this volume are cross-cutting connections and deepening lines of inquiry into some core areas of theoretical study that are crucial for grasping the challenge of adapting humanity.

We have already mentioned how the volume approaches adaptation and human flourishing through four specific contexts: the role of history in ecological restoration under climate change, the status of the environment in the virtue of justice (or in affiliated virtues) as seen through the promise of the capabilities approach, the extent to which global climate change ought to change the content or form of the character virtues, and the vexed relationship between virtue and its wider social and political contexts. In closing this introduction, we outline how these four contexts actually speak to each other and reinforce a deepening inquiry. We chart one way the argument of the volume proceeds across its entirety.

In chapter 1 Ned Hettinger explores how restoration has been proposed as a key to developing our relationship with our planetary environment. While he concludes that it should not be paradigmatic of that relationship, Hettinger and other contributors in part I all accept that ecological restoration can offer valuable insights into how we should live well in relation to our environment under global climate change. This opens discussion about the nature and substance of restoration's insights when they are scaled to a planetary level.

Part I's debate about historical fidelity in ecological restoration is a debate about how our knowledge and expectations of the environment should respond to ecological change that appears likely to outstrip what Eric Higgs in chapter 4 calls our "ecological imagination." At the heart of this debate is also the question of how much we ought to manage the environment when we act as stewards of our relationship to it. In effect, part of what the debate around historical fidelity traces is the contour of the environmentally respectful yet practically realistic constraints on our stewardship of the environment (Sandler, chapter 3, this volume). Moreover, the debate does so by raising how these constraints ought to show up in a life well lived, that is, in the virtues we expect of humane people in a new global climate in which environmental stewardship is an inevitable consequence of recognizing environmental value (see, especially, Throop, chapter 2, this volume).

But once we see how our stewardship of the environment ought to change in recognition of a plurality of values and accept that these changes ought to be mirrored in shifts within human virtue, we face pressing questions about how this should affect core human virtues, such as courage, temperance, wisdom, and justice. For example, in chapter 5, Light's emphasis on value pluralism in setting restoration goals makes his work open to the role of human capabilities in guiding restoration, as Holland advocates in chapter 7. To pursue this line of inquiry, chapters in part II consider justice because it is a macrovirtue that affects every other virtue. Justice is a key virtue of relationships and we exist in relation to each other and to our environment, both of which are ineliminable facts about human life on Earth. Thus, discussion of justice exemplifies an important investigation into the depth to which our altered relation to the environment, first examined in part I, ought to permeate our character. And the answer is surprising: it ought to permeate our character and our institutions in profound and far-reaching ways (Holland, chapter 7, this volume; Schlosberg, chapter 8, this volume).

Among the many different theories of justice, one may ask why capability theory—the theory developed by Amyarta Sen and Martha Nussbaum (Nussbaum 2000, 2006), institutionalized within the UN World Human Development Reports, and now the subject of a sizable international association of scholars, policymakers, and practitioners—is especially well poised to help us understand the relevant shift in the role that justice plays in our lives. Keulartz and Swart, Holland, and Schlosberg all address this question in their chapters. The general idea they share is that *capability theory is conceived in a protoecological way already*. Its concepts are especially well disposed for extension or application to our environment (Keulartz and Swart, chapter 6, this volume). And so part II's specific context, far from being an in-house debate about capability theory, now has general relevance to thought about human flourishing under significant environmental change.

Leading into part III, then, we have seen a growing case for why and how human flourishing ought to change under global climate change. We have seen, first, how it ought to change in our stewardship of the environment, and second, how it ought to change when we consider justice *within* that relationship. We are now well underway in our pursuit of the idea of adapting humanity. What remains to be seen is how revising aspects of our moral self-conception should actually play out. And this is where attention to the question of who we are and a focus on specific character traits, their form, and their content come in.

Taking up Dale Jamieson's advocacy in chapter 9 for an adjusted understanding of the virtues as an essential feature of our moral response to climate change, the chapter authors in part III consider some fundamental dimensions of how character virtue ought to be adjusted. Thompson in chapter 10 considers a concept of responsibility *through which* any virtues of the future will have to pass or *with which* the virtues of stewardship will have to be coordinated. Kawall's study in chapter 11 of the vice of greed then explores the way that the scope and cause of climate change demands a fairly radical reconceptualization of what counts as vice and, further, how vice (and by extension virtue) ought to shift to internalize human responsibility for the scale on which climate change and its causes operate. We believe these kind of ideas are prefigured but undeveloped in Jamieson's seminal work, reprinted here as chapter 9. Then Gardiner in chapter 12 presents a case study at the limit of responsibility gone wrong, that is, the viciousness of over-management and domination already prefigured in part I, in the chapters of Hettinger, Throop, and Sandler. Gardiner shows us what is at stake, morally, if we succumb to that vice of excess, indulging a longing for control that misapplies Thompson's virtue of global responsibility.

Already Kawall and Gardiner in their research begin to explore the pressures that an institutional context places on virtue. Kawall's "moderate greed" (a vice too many of us exhibit) is sheltered and promoted by large-scale institutional arrangements, and Gardiner's argument about the moral dangers of geoengineering circles around difficulties connected with our civic and political responsibility. Accordingly, part IV takes seriously the institutional challenges to virtue, developing a series of problems that any further research about adjusting virtue, justice, and our stewardship of nature would do well to consider.

First, virtues of character will have to be put in their social, political, and biological contexts. Humanity must take responsibility not only for the planetary environment that we have brought about, and the attendant loss of environmental value we cherish, but for the institutions and practices that mediate our relations to the environment (Bendik-Keymer, chapter 13, this volume). Moreover, we have to face our evolutionary heritage. Just as Higgs notes in chapter 4 that we are likely to lack the ecological imagination for the new world climate—having evolved and recorded all of our history thus far in an environment dramatically different than the one predicted for 2100—so too we are likely to have cognitive or other adaptive difficulties resulting from our evolutionary heritage. Today, we face problems on a scale that the human species has never faced before. Hirsch and Norton explore some key aspects of adapting our institutions to an appropriate scale in chapter 16.

Second, the institutional contexts in which we try to adapt ourselves must be seen with all the internal limits they contain. Thus, in chapters 14 and 15 Shockley and Vogel each confront us with some of the difficulties involved with embedding virtue in an institutional context. Not only might the underlying psychology have already suffered distortion from the beliefs and norms reinforced by the context (Vogel, chapter 15, this volume), but to echo Robert Jackall (2010), the operational structure of institutions may also disorganize virtue (Shockley, chapter 14, this volume).

In sum, the research collected in this volume suggests that adequately adapting humanity to global climate change involves pursuing further work on character and context that moves back and forth between the incomplete insights of ecology, the revised demands of environmentally informed justice, the reframed form and content of character virtues keyed to a planetary scale, and the vexing internal dynamics of the organizational inertia of our civilization.

Notes

1. At the time of writing, there is still no ambitious, multilateral mitigation agreement in place that is capable of delivering a 50 percent cut in global greenhouse gas emissions by 2050, giving us a reasonable chance to limit warming to +2°C. The unilateral commitments under the Copenhagen Accord provide only two thirds of the total reductions needed by 2020, in order to be on the 2050 mitigation pathway. We thank Andrew Light (2010) for this clarification.

2. See the outcome of the Ad Hoc Working Group for Long Term Cooperative Action, http://unfccc.int/meetings/cop_16/items/5571.php (accessed February 15, 2010).

3. The concept of adaptation is also used in the natural sciences, especially in population biology and evolutionary ecology. The concept was subsequently adopted by the social sciences, where other ecological concepts such as tolerance, stability, and resilience are currently used to discuss human interactions with their environment. See Smithers and Smit (1997) 2009. An overview of adapta-

tion in the UN's response to climate change can be found in Schipper (2006) 2009. An overview of research on vulnerability and adaptation to climate change in social and climate science can be found in Leichenko, Thomas, and Barnes 2010.

4. Vulnerability to climate change is broadly defined as "the degree to which an individual, group, region or system is susceptible to and is unable to cope with adverse affects of climate change." The concept has a long history in social sciences relating to natural disasters and food security (Leichenko, Thomas, and Barnes 2010, 135).

5. Specific priorities include (1) defining "vulnerability" (should, for example, only the Least Developed Countries (LDCs) and Small Island Developing States (SIDSs) count as vulnerable or should *all* developing nations count as vulnerable?); (2) setting baselines for so-called new and additional adaptation funding (is good adaptation funding nothing more than good human development funding?), and (3) determining compensation for unavoidable losses (for example, who ought to pay for the destruction wrought by episodic weather events of increased intensity?). See, for example, Adger et al. 2006 and Baer 2006.

6. Some relevant questions here include: How should climate funds be allocated between mitigation and adaptation? Should adaptation funds be administered by new or existing institutions? Should these institutions be operating inside or outside the UNFCCC? Should the financing consist of public funds exclusively or could it include private sources and loans?

7. The Copenhagen Accord and the Cancun Agreements require adaptation funding to be "new and additional" to existing funds earmarked for economic development projects. Yet the *goal* of that adaptation funding still is to allow the country to maintain development traditionally understood as correlated with economic productivity (see UNFCCC Article 2 and notes 8 and 12).

8. Reducing vulnerability also includes increasing resilience.

9. Our discussion is focused on how the concept of adaptation operates at the level of international climate policy and so represents a top-down approach, framed in terms of welfare economics. We do not intend to deny that climate change can threaten very real, immediate, and more palpable harms, such as food security and access to drinking water. Most community-based, bottom-up approaches to adaptation are aimed at the relief or prevention of such suffering caused by climate change to the most vulnerable people. By criticizing how adaptation is conceptualized at the level of international policy we do not mean to suggest that efforts to promote adaptation at the local level, understood as reducing vulnerability to climate harms, are ill conceived.

10. This applies whether or not they are, to use Jamieson's (2005) terminology, "reactive" or "anticipatory."

11. We are *politikon zoon*—political animals—as Aristotle ([ca. 350 BCE] 2002) said.

12. This raises issues about the difficulty of defining restoration and the relevant standards for success, topics Light treats in chapter 5 of this volume.

References

Adger, W. N., J. Paavola, S. Huq, and M. J. Mace. 2006. *Fairness in Adaptation to Climate Change*. Cambridge, MA: MIT Press.

Aristotle. (ca. 350 BCE) 2002. *Nicomachean Ethics*, trans. Christopher Rowe. New York: Oxford University Press.

Aristotle. ca. 350 BCE. *Politics*. Many editions available.

Baer, Paul. 2006. Adaptation to Climate Change: Who Pays Whom? In *Fairness in Adaptation to Climate Change*, ed. W. N. Adger, J. Paavola, S. Huq, and M. J. Mace, 131–155. Cambridge, MA: MIT Press.

Burton, I. 2009. Climate Change and Adaptation Deficit. In *Adaptation to Climate Change*, ed. E. L. F. Schipper and I. Burton, 89–99. London: Earthscan.

Clewell, A., J. Aronson, and K. Winterhalder. 2004. "The SER International Primer on Ecological Restoration." Society for Ecological Restoration International Science & Policy Working Group. http://www.ser.org/content/ecological_restoration_primer.asp#3 (accessed March 25, 2009).

Constant, B. 1819. *On the Liberty of the Ancients Compared with That of the Moderns*. Many editions available.

Crutzen, P. J. 2002. Geology of Mankind. *Nature* 415: 23. http://www.nature.com/nature/journal/v415/n6867/full/415023a.html (accessed March 9, 2009).

Davies, S. 2009. Are Coping Strategies a Cop-Out? In *Adaptation to Climate Change*, ed. E. L. F. Schipper and I. Burton, 99–116. London: Earthscan.

Gardiner, Stephen. 2011. *A Perfect Moral Storm: The Ethical Tragedy of Climate Change*. New York: Oxford University Press.

Gardiner, Stephen, Simon Caney, Dale Jamieson, and Henry Shue, eds. 2010. *Climate Ethics: Essential Readings*. New York: Oxford University Press.

Hamilton, Clive. 2010. *Requiem for a Species: Why We Resist the Truth about Climate Change*. London: Earthscan.

Hegel, G. W. F. 1820. *Elements of the Philosophy of Right*. Many editions available.

Hobbs, Richard J., et al. 2006. Novel Ecosystems: Theoretical and Management Aspects of the New Ecological World Order. *Global Ecology and Biogeography* 15: 1–7.

IPCC. 1996. *Climate Change 1995: Impacts, Adaptations and Mitigation of Climate Change: Scientific-Technical Analyses*. Report of Working Group II. Contribution of Working Group II to the Second Assessment of the Intergovernmental Panel on Climate Change, ed. R. T. Watson, M. C. Zinyowera, and R. H. Moss. Cambridge, UK: Cambridge University Press.

Jackall, Robert. 2010. *Moral Mazes: The World of Corporate Managers*. 2nd ed. New York: Oxford University Press.

Jamieson, Dale. 1992. Ethics, Public Policy, and Global Warming. *Science, Technology and Human Values* 17 (2): 139–153.

Jamieson, Dale. 2005. Adaptation, Mitigation, and Justice. In *Perspectives on Climate Change*, ed. W. Sinnott-Armstrong and R. Howarth, 221–253. Amsterdam: Elsevier.

Kelly, P. M. and W. N. Adger. 2009. Theory and Practice in Assessing Vulnerability to Climate Change and Facilitating Adaptation. In *Adaptation to Climate Change*, ed. E. L. F. Schipper and I. Burton, 161–186. London: Earthscan.

Leichenko, R., A. Thomas, and M. Barnes. 2010. Vulnerability and Adaptation to Climate Change. In *Routledge Handbook of Climate Change and Society*, ed. C. Lever-Tracy, 133–151. New York: Routledge.

Leopold, A. 1949. *A Sand County Almanac, and Sketches from Here and There*. New York: Oxford University Press.

Lever-Tracy, C., ed. 2010. *Routledge Handbook of Climate Change and Society*. New York: Routledge.

Light, Andrew. 2010. "So Close, Yet So Far." Center for American Progress. http://www.americanprogress.org/issues/2010/12/so_close.html (accessed July 20, 2010).

Lovelock, J. 2006. *The Revenge of Gaia: Earth's Climate Crisis and the Fate of Humanity*. New York: Basic Books.

Mill, J. S. 1859. *On Liberty*. Many editions available.

Moore, Frances. 2008. "Carbon Dioxide Emissions Accelerating Rapidly." *Earth Policy Institute*. http://www.earth-policy.org/indicators/Temp/carbon_emissions_2008 (accessed August 21, 2011).

Nordhaus, W. D. 1994. *Managing the Global Commons: The Economics of Climate Change*. Cambridge, MA: MIT Press.

Nussbaum, M. 2000. *Women and Human Development: The Capabilities Approach*. New York: Cambridge University Press.

Nussbaum, M. 2006. *Frontiers of Justice: Disability, Nationality, Species Membership*. Cambridge, MA: Belknap Press of Harvard University Press.

Pew Center. 2007. Climate Change 10: Understanding and Responding to Global Climate Change—Adaptation. http://www.pewclimate.org/docUploads/Climate101-Adaptation-Jan09.pdf (accessed June 29, 2011).

Plato. ca. 380 BCE. *The Republic*. Many editions available.

Rousseau, J.-J. 1762a. *Émile, or on Education*. Many editions available.

Rousseau, J.-J. 1762b. *The Social Contract*. Many editions available.

Sandler, R. 2007. *Character and Environment: A Virtue-Oriented Approach to Environmental Ethics*. New York: Columbia University Press.

Sandler, R., and P. Cafaro, eds. 2005. *Environmental Virtue Ethics*. Lanham, MD: Rowman & Littlefield.

Schipper, E. Lisa F. (2006) 2009. Conceptual History of Adaptation in the U.N.F.C.C.C. Process. In *Adaptation to Climate Change*, ed. E. L. F. Schipper and I. Burton, 359–377. London: Earthscan.

Schipper, E. Lisa F., and I. Burton, eds. 2009. *Adaptation to Climate Change*. London: Earthscan.

Shue, H. 1999. Global Environment and International Inequality. *International Affairs* 75 (3): 531–545.

Smithers, J., and B. Smit. (1997) 2009. Human Adaptation to Climatic Variability and Change. In *Adaptation to Climate Change*, ed. E. L. F. Schipper and I. Burton, 15–33. London: Earthscan.

Steffen, A., ed. 2006. *Worldchanging: A User's Guide to the 21st Century*. New York: Abrams.

I

Adapting Restoration to Climate Change

1
Nature Restoration as a Paradigm for the Human Relationship with Nature

Ned Hettinger

Throughout most of the twentieth century, preservation was the reigning nature-protection paradigm. On this view, protecting nature involves setting aside nature preserves and keeping them "untrammeled by man" (U.S. Congress 1964). For preservationists, nature's key value is its "naturalness" or "wildness," that is, the degree to which it is independent of human influence. According to this paradigm, humans are, by and large, separate from nature and human involvement with nature degrades it. Among the virtues preservationism promotes are moderation, humility, and fairness.

In the last quarter century, however, nature restoration has become the major competing paradigm for the protection of nature. Given increasing human alteration and degradation of nature and greater awareness and understanding of these effects—a problem made increasingly poignant by anthropogenic climate change—attempts to restore degraded nature have become a key environmental goal. Examples include a controversial but successful restoration of gray wolves to the Greater Yellowstone ecosystem and an ongoing multibillion-dollar attempt to restore parts of the Everglades. Fire is being returned to—rather than banished from—many fire dependent ecosystems. Dams are being removed for the first time in history, and there are numerous campaigns to remove human-introduced, invasive, exotic species.

The restoration paradigm has been articulated by way of challenges to preservationism. Restorationists claim that preserving nature won't save it; instead, we must restore nature if it is to continue. Among other factors, they point to damage to nature caused by global climate change as indicating the necessity of restoration for nature protection. Rather than locating nature's value in its lack of humanization (as do preservationists), restorationists see nature's value in its thriving biodiversity. Restorationists reject the conception of humanity as separate from nature and argue that restoration is a virtuous way for humans to be part of nature. On their view, restoration is necessary both for nature and for human flourishing. Instead of moderation, humility, and fairness, restorationists stress the virtues of engagement, competency, and taking responsibility.

Important environmental figures sing the praises of restoration. For instance, MacArthur fellow Gary Nabhan has written: "The emergence of ecological restoration is . . . the most important environmental development since the first Earth Day. It allows people to participate in healing the wounds left on the earth, acknowledging the human power to create as well as to destroy" (1991, 4).

And writer Michael Pollan (2003) claims that "Ecological restoration is one of today's most constructive, hopeful, and provocative environmental movements." He identifies William Jordan III as "its leading visionary."[1]

Yet there have been many critics of restoration and the restoration paradigm, and environmental philosophers have been among the most vocal. Robert Elliot (1982, 1997), in his well-known paper and book *Faking Nature*, worries that restoration can be used to undermine preservation. He argues that if a restored nature were considered as good as original nature, it would be irrational to preserve nature rather than utilize/degrade nature and then restore it. Elliot rejects the "restoration thesis" and argues that a restored nature is not equally good as the original; instead, it is fake nature.[2] Like a replicated artwork, it is not as valuable as the original for it lacks the type of genesis that provides an important reason for valuing it. Rather than being a product of natural history, a restored nature is a product of human culture and technology.

Stanley Kane has argued that the restoration paradigm amounts to a paternalistic domination of nature: "By holding that humans are the lords of creation, restorationist metaphysics tolerates no enclaves anywhere kept free of human domination and control" (1994, 83). The restorationist manipulates and controls nature for its own good, deciding, for example, when nature will burn and what plants and animals are to be allowed.

Eric Katz, the dean of the anti-restorationists, joins Kane's view with Elliot's concern with inauthentic nature. Katz sees restoration as "The Big Lie":[3] "A 'restored' nature is an artifact created to meet human satisfactions and interests," he states. "It is an unrecognized manifestation of the insidious dream of the human domination of nature. . . . Humanity will demonstrate its mastery of nature by 'restoring' and repairing the degraded ecosystems of the biosphere" ([1992] 1997, 95).

Katz claims that restored "nature" is an anthropocentric human artifact and "nothing more" (Katz 2007, 103). When we restore, he says, "we are creating artifactual systems–or at best, hybrid systems of natural entities and artifacts" (2000, 39) that resemble nature but are not authentic nature. Rather than healing nature and making it whole and healthy again, restoration is "putting a piece of furniture over the stain in the carpet" ([1992] 1997, 106).[4]

Should we accept nature restoration as the new environmental paradigm for a virtuous human relationship with nature and give up on the old environmental paradigm of preservation? Or instead, should we embrace the critics' contention that the restoration paradigm is deeply flawed and pernicious? I believe that ambivalence toward restoration and the restoration paradigm is most appropriate. The restoration paradigm presents serious challenges to preservationism and identifies important components of a healthy human relation to nature. But the restoration paradigm also involves deep confusions. Advocates of restoration are good at diagnosing problems in preservationism and criticizing its excesses, but their solutions fail to provide for what they show us we need. This much is clear: The debate between the restoration's proponents and its critics teaches us much about how we should conceive of the human/nature relation.

Strengths of the Restoration Paradigm

There is much of value in the practice of restoration and the restoration paradigm. I shall consider four points.

1   Restoration Can Help Nature

Restoration is an important and valuable human activity, and it need not be anthropocentric in motive or result. Not only can restoring degraded nature help humans, but also restoration has the power to help heal nature and let a degraded piece of nature once again flourish on its own. For example, the restoration of wolves to the Greater Yellowstone ecosystem is not just good for us (e.g., providing more exciting tourist visits and increased income for park-related businesses), but it also has made the gray wolf species more robust and it arguably has improved the health and integrity of the Yellowstone ecosystem. In many cases, restoration is needed for the sake of nature.

One might think that any thoughtful commentator on restoration would acknowledge this point. After all, even Eric Katz favors restoration: "Nothing I have said . . . should be taken as an endorsement of actions that . . . injure areas of the natural environment and leave them in a damaged state" ([1992] 1997, 105–106). "I believe that remediation of damaged ecosystems is a better policy than letting the blighted landscape remain as it is" (2002, 142).

But Katz can only endorse *anthropocentric* restoration, that is, restoration projects whose motive and result are for the benefit of humans. He cannot support restoration for nature's sake. On Katz's view, restoration creates artifacts or further artifactualizes ecosystems already affected—and thus artifactualized—by humans. Therefore for the sake of nature's value, the best we can do is to leave degraded nature alone. But it seems clear that restoration is needed not just for the benefit of humans but for nature as well.[5] Restorationists helpfully insist on this point.

2   Hands-off Preservation Is Not Enough

The restoration paradigm acknowledges the massive damage humans have caused (and continue to cause) the natural world. Preservationism ignores the extent to which human influences on nature are back-loaded and pretends that nature will be okay if we just leave it alone. But sometimes, as Jared Diamond (1992) insists, "We must shoot deer to save nature." Sometimes, human inaction can mean the further degradation of natural areas due to ongoing affects of past human action. Left alone, exotics introduced by humans can wreak havoc on native ecosystems, species, and individuals. It is arguable that, after decades of fire suppression, unless humans actively reduce fuel loads, fires on a scale that seldom would have occurred naturally will denude huge swaths of the landscape. Restoring predators to ecosystems from which they were extirpated can stop the ongoing degradation due to the overabundance of prey. Restoration recognizes the ongoing harm we have caused and seeks to make amends.

But this idea can be taken too far. Advocates of the restoration paradigm often overstate the necessity of restoration. For example, William Jordan, the founder of the journal *Ecological Restoration* and perhaps the most forceful proponent of the restoration paradigm, claims we will need to restore and manage the entire earth. "Preservation," he says, "is impossible. . . . All systems are constantly changing, and . . . this change reflects at least some degree of human influence" (1994, 19). The

Midwest's tall grass prairies and oak openings are examples where, Jordan says, "The entire native ecosystem has been virtually eliminated as a direct or indirect result of new kinds of human activities. . . . This situation is actually paradigmatic, however, and is true in the final analysis of all ecosystems everywhere" (19–20).

But it is not true that all human influence requires or justifies restoration. That humans have affected virtually the entire surface of the planet doesn't make preservation without restoration impossible. For example, a slightly higher level of acidity in Yellowstone's rain does not make restoration necessary. Nor does it mean that we are preserving an unnatural Yellowstone. Only a perverse fixation on absolute purity would lead to these conclusions.

One might argue that human-induced global climate change justifies the contention that we will need to restore the entire earth, especially if we want it to become natural again. For changing the climate in a region will have impacts on all the biota that live there and on much of the abiota as well. Such influence leads Allen Thompson (chapter 10, this volume) to conclude that humans are now accountable for the basic conditions of life on earth. Bill McKibben has famously argued that such massive human influence means that we are at the "end of nature": "*We* ended the natural atmosphere, and hence the natural climate, and hence the natural boundaries of the forests, and so on. . . . (1989, 78) There is no such thing as nature any more (89). As we heated the planet . . . we would change the flora and fauna *everywhere;* even at the poles or in the Adirondack wilderness, we now influence every physical system (1995, 10). Nature as something separate from man has vanished" (11).

For a variety of reasons I don't think global climate change has the conceptual power that McKibben in his early writing gives it. Nor do I think that it provides a justification for the claim that we need universal restoration or that it is desirable.[6] Here again a dominant motivation is a problematic desire for an absolutely pristine nature and the even more problematic insistence that real nature must be a virginal nature—untouched by man. But nature can be nature and remain natural even while impacted by humans to a significant extent. There are degrees of naturalness and being somewhat less natural does not entail being unnatural or losing the status of nature.[7]

Further, although caused by humans, the dramatic changes in nature that have resulted and will continue to result from climate change do not constitute human domination or control of nature. We neither intend nor control the consequences for nature and humans that climate change brings. Moreover, climate change makes us more vulnerable to nature and less able to predict or control it. With climate change, nature is very much in charge, even though humans are responsible for pushing natural systems onto different trajectories. For these reasons, I do not think climate change has brought about (or will bring about) the end of nature or has placed us into, as McKibben has suggested, "a world of our own making" (1989, 85).

I do not belittle the harm that climate change can and is causing to humans, animals, plants, species, or ecosystems (and even to the naturalness of nature).[8] And I believe that restoration activities are an important response to this harm, including restoration with a strong emphasis on historical fidelity.[9] When warming ocean temperatures bleach coral reefs and subsequently kill them, humans are harmed (by loss of tourism and fisheries), habitat for fish species is destroyed, and spectacularly beautiful and amazing animals and ecosystems are wiped out. This is the end of very special individual pieces of nature. In so far as climate change threatens massive die offs of species and ecosystems, humans have responsibilities to try to mitigate and prevent these harms, and restoration must play a role. But I don't think it plausible to claim that virtually all species and ecosystems will need to rely on human intervention and restoration for their survival. Species and ecosystems can and do migrate as climate shifts, although the rapidity of human-caused climate change is a serious worry here. Some ecosystems and species are tolerant of a wide range of climate conditions. Moreover, it is likely that many of these species and ecosystems will adapt or evolve in response to a changing climate and the goal of preventing this adaptation or evolution, or recreating the previous climate regime—perhaps via geoengineering (see Gardiner, chapter 12, this volume)—will often result in much more heavy-handed human manipulation and management of nature that is involved in human-caused climate change. Nevertheless, restorationists are right that ongoing degrading effects of human impacts on nature require efforts at restoration and that a hands-off preservation policy is not sufficient.

3   Full Human Participation

A third insight of the restoration paradigm is that it insists on full human participation in nature. William Jordan has argued that preservation offers a severely limited human relation to nature. It limits people's role in nature to a nonparticipatory "take only pictures and leave only footprints." It makes humans visitors on the planet, instead of active, contributing members. The restoration paradigm sees active human participation in nature as a necessary part of a healthy human/nature relation and views the self-abnegation in preservationism as incompatible with human flourishing.

Eric Katz defends such a nonparticipatory approach to nature when he writes:

Here is my solution: as much as possible, we humans *leave nature alone.* To "let it be" seems to be to be the highest form of respect we can muster. . . . And while I leave it alone, I try to learn as much as possible about it, so that knowledge, respect, and love can all grow together. We can use the art object/nature analogy again. . . . If I respect a work of art, I show this respect by my mere appreciation, by learning about the artwork . . . I do not attempt to change the work of art . . . I do not attempt to improve it . . . Any intervention in the artwork itself will change its quality and value. My proper respectful role is to leave the physical object alone. (2002, 143)

For a "pure" preservationist like Katz, appropriate respect for nature is like appropriate respect for art: One should appreciate it and leave it alone.

But surely a healthy relationship with nature must involve more than this. I think Jordan is right that such a relation "must engage all our abilities . . . our physical, mental, emotional and spiritual capacities" (1994, 18–19). It should be a "working relationship" including "ecological interaction." Preserving wildland that we study, love, and leave alone is not the only dimension of a healthy human relationship with nature, though it certainly is one of them.

Elliot rejects the restoration paradigm's favorable attitude toward restored ecosystems that require continued human manipulation and management. The "ultimate aim implicit in the preservationist ideal is to achieve a situation from which humans are absent, except as respectful, careful and unobtrusive visitors" (Elliot 1997, 145). Defending the restoration paradigm in response, Steve Vogel argues: "Such an aim seems to me to express a deep alienation from nature and a failure to understand the human role in it. We are not visitors on Earth, and indeed we are never absent from it—not, nowadays, from any of it. Restoration's value, I think, would rather come precisely from our experience in it of our involvement in the world, our responsibilities regarding it" (2003, 165).

Visitor status for humans in designated wilderness areas is a good model. But Vogel is right that such a status is not the right way to think of humans' relation to the entire planet.[10] We need a more active, participatory, involved conception of the human role in earthen nature than we get from modeling humans' relation to nature on how we should relate to wilderness areas or art objects.

We might reinforce this restorationist critique of preservation by calling attention to what might be called a tendency in preservationism toward human/nature apartheid. The belief that respect for nature requires such separation is based on the dubious idea that, as one commentator expresses it, "nature can be fully itself and thus have full value only when left undisturbed by humans" (Kane 1994, 71).[11] Philosopher John Visvader puts his finger on the problem when he argues "We need to understand both the 'natural' and the 'wild' in such a way that we can imagine giving more to the world around us than the gift of our mere absence" (1996, 18).

4   Positive Role for Humans

Finally, the restoration paradigm sees the need for a positive role for humans in nature. Restorationists criticize preservationism for lacking such a vision.[12] Insofar as it has a model for human's place in nature, preservationism would seem to embrace primitivism. On this view, benign human participation in nature requires that we "go back to the Pleistocene" and adopt a hunter-gatherer lifestyle.[13] Restorationists have pointed out that this conception of appropriate human community with nature would have us give up much of what makes for human flourishing (namely, the ongoing development of civilization and technology).

Restoration offers itself as the missing positive role for humans in nature. Restoring nature involves the active use of science and technology and thus, unlike preservationism's primitivist model, the restoration paradigm does not require repudiating these key achievements of civilization.[14] Jordan argues that restoration involves the "re-inhabitation of nature" "without abandoning the lessons learned on 'the pathway to the moon'" (1994, 23).[15]

Jordan and the restorationists are right to seek a positive, active role for humans in nature, one that uses the full range of human abilities. But this insight is tarnished when Jordan argues that this reinhabitation of nature can occur without giving up what he calls "the accouterments of civilization" (1994, 21). Jordan apparently thinks that the accessories, equipment, and furnishings of modern-day living do not get in the way of a positive role for humans in nature. But there will be no healthy human-nature community without reducing consumption and abandoning our environmentally unfriendly technologies and ways of life.

Weaknesses of the Restoration Paradigm

While there is much to learn from the restoration paradigm and particularly from its critique of pure preservationism, there are also serious problems with taking restoration as a positive model for human/nature relations. I will discuss four.

1   Restoration as Grandiose and Hubristic

First, the restoration paradigm tends to be grandiose and hubristic. William Jordan's description of the importance of restoration is particularly grandiose. "Restoration has the elements of a kind of ritual, even a sacrament, of reentry into nature' (1986, 25), he writes. "It's a way of participating in the Creation with a capital 'C'" (2002, 27). Furthermore: "While the preservationist in us continues to believe, with Columbus, that Eden actually exists . . . the restorationist has turned to a different task—the task not of finding an existing Eden, but of actually making it out of the raw materials in a landscape compromised by history" (1992, 3).

The hubris in restorationism can be seen in its contention that "nature needs us" in some fundamental way. Chicago environmentalist and noted restorationist Stephen Packard once suggested that those who restore should be seen as "parents" of nature. He writes: "It's an honor to be among the first to have a nurturing relationship with wild nature. . . . If we are dependent on nature, what's so terrible about nature being dependent on us too. . . . In some ways nature was our parents and now we're its parents. Now it depends on us" (1990, 72). Packard also notes, "A restorationist, like a parent, need[s] to protect an unsteady being from certain insults to its health or existence . . . [and] help some life go forward on its own" (1993, 14).

I find it presumptive to think that those involved in the contemporary restoration movement are trailblazers in having a nurturing relation with (wild) nature. One would think, for example, that native peoples—who saw their relation and use of nature in spiritual terms—had (and have) a nurturing relationship.[16] Traditional methods of farming and gardening might also justifiably be seen as nurturing. Or consider the practice of feeding birds or the activity of those individuals, like Johnny Appleseed, who have helped particular species flourish in new habitats. Additionally, the violence toward existing flora and fauna involved in many restoration projects (including those that Packard has supervised) gives lie to the notion that restoration is nurturing in its relationship with nature.[17]

Jordan argues that "shame" is necessary in our relationship with nature: "Restoration is shameful because it involves killing and a measure of hegemony over the land" (2003, 50). Those who have a nurturing relationship with another are not easily seen as engaged in shameful behavior toward them.

The suggestion that restorationists should be seen as parents of a child-like nature is equally presumptuous. Nature on earth really did produce human beings and continues to provide absolutely essential conditions and resources for human life. While it is true that restoration can return ecosystem processes that were absent in an area and bring back extirpated species, a healing metaphor is much more apt and far less tendentious than the model of parenting.[18] Even in cases where restoration results in new ecosystems in an area, such systems are not produced from scratch, but rather use preexisting ingredients. Flora and fauna are moved in, not constructed. The soil is rearranged, not produced. In restorations involving removal of exotics or human structures, or restorations that involve cleaning up harmful chemicals or trash, a janitorial model seems best.

Holmes Rolston's response to Packard's idea of humans as parents of nature is helpful: "The parent-child analogy is misleading," he writes. "Parents cease to operate as parents when they are dependent on us. Though, owing to the inevitable decline of individuals, parents will become dependent on their children, we do not want to cultivate those dependencies. Our parents are failing when these are required. Nature is not some failing parent that now needs to become dependent on us" (1994, 201).

It is misleading and arrogant to think of nature as old and worn out, as no longer able to provide for us, and as now needing us to take care of it. Further, to the extent it is true that some elements of nature have become dependent on us for their continued existence, this is not something to be celebrated. In many cases, what these pieces of nature need from us is to stop attacking them and to leave them alone. There remains a good deal of truth in the preservation perspective. Far from needing us, much of nature would be far better off without humans on the planet.[19] Even in cases where active human restoration is required (e.g., with endangered species such as the California condor or the red wolf), humans act much more like physicians who transplant organs than parents who give birth.

Part of what is objectionable is the suggestion that nature *as a whole* has become dependent on our restoration activities, rather than particular elements and forms of nature being so dependent. Jordan claims that: "It seems obvious that . . . the fate and well-being of the biosphere depends ultimately on us" (1994, 27). Stephen Jay Gould's response to the often-heard suggestion that humans must save the planet points out the scientific ignorance and moral failings involved in this suggestion. He writes:

> Such views . . . are rooted in the old sin of pride and exaggerated self-importance. We are one among millions of species, *stewards of nothing*. By what argument could we, arising just a geological microsecond ago, become responsible for the affairs of a world 4.5 billion years old, teeming with life that has been evolving and diversifying for at least three-quarters of that immense span? . . . We are virtually powerless over the earth at our planet's own geological time scale. All the megatonnage in our nuclear arsenals yields but one ten-thousandth the power of the asteroid that might have triggered the Cretaceous mass extinction. Yet the earth survived that larger shock . . . [which] paved the road for the evolution of large mammals, including humans. We fear global warming, yet even the most radical models yields an earth far cooler than many happy and prosperous times of a prehuman past. We can surely destroy ourselves, and take many other species with us, but we can barely dent bacterial diversity and will surely not remove many million species of insects and mites. On geological scales, our planet will take good care of itself. . . . Our planet simply waits. (Gould 1990, 217)

The picture of the restorationist as a nurturing parent creating Eden out of a needy nature is grandiose and arrogant.[20]

2   Ignores the Value of Wildness

A second major problem with restorationism is that it is insufficiently appreciative of the value of wildness in nature. For the most part, restorationists seem blind (or openly hostile) to the value of nature as other, to having a world that is not of our own making, and to the importance of minimizing human impacts on nature. I believe that any appropriate appreciation of nature and any proper conceptualization of the human role in nature must place significant evaluative weight on naturalness and particularly on nature's autonomy, that is, the degree to which a natural entity is not dominated or controlled by humans.[21]

Jordan hopefully predicts that "restoration will become the principal outdoor activity of [the] next century and the result will be the conversion of nature from . . . 'environment' into habitat for human beings" (1994, 23). One wonders if this includes what is commonly thought of as wild nature, namely, national parks, wilderness areas, mountains, and deserts. Restorationists see no problem at all with, as Jordan puts it, "leaving a distinctively human mark on the landscape" (17). Ongoing human management of restored landscapes is not seen as a problem but as a positive opportunity for human involvement in nature. On the restoration paradigm, it seems, nothing is off limits, as long as humans are helping to restore degraded nature. This perspective ignores the value of having some earthen biotic nature free from human control and manipulation.

Jordan's use of a garden metaphor clearly illustrates this insensitivity to the value of wildness in nature. Jordan has characterized ideal nature as a type of human garden. He argues: "Whether we wish to admit it or not, the world really is a garden, and invites and even requires our constant participation and habituation" (1986, 25). "Restoration," he writes, "is that form of gardening concerned specifically with gardening, maintenance, and reconstitution of wild nature and is the key to a healthy relationship with it" (1994, 18). Appreciation of the value of nature as other and respect for nature's autonomy are not compatible with conceiving or treating nature as if it were significantly like a human garden. (In his later writing Jordan explicitly takes back this unfortunate metaphor.)[22]

3   Restoration Not a Net Benefit

A third major problem with the restoration paradigm is that it misconceives restoration as providing a net-benefit to nature, instead of an attempt to heal or engage in restitution for harms caused. Jordan claims, "ecological restoration provides a basis—actually a paradigm—for a healthy, mutually beneficial relationship between ourselves and the natural landscape" (1994, 18). Jordan sees restoration as a human gift to nature. It is, he says, "Our gift back to nature. The restored ecosystem is something that we offer nature in return for what nature has given us. . . . It represents what is in a sense our best gift" (Jordan 2000, 25).[23] But to degrade some natural entity or system and then attempt to restore it (even successfully) is clearly not to benefit nature or give it a gift. Instead, restoration is restitution for past harm and cleaning up of our mess. When a wife-beater gives his victim first aid, it is not a "gift" or net benefit. When an oil "spill" soaks beaches, cleaning it up is not a gift or net benefit to nature. When we restore wolves to an ecosystem from which we eradicated them, this is an attempt to make amends for past wrongs, to put back what we have taken away. Concepts like "gift," "benefit," and "exchange" do not help us understand what we are doing when we restore a nature that we have degraded.

It is often true that individual humans who help restore a piece of nature may not have been directly involved in its destruction. Thus one might be tempted to claim they are providing a benefit or giving a gift to nature. But given that virtually all of us who live in modern industrial societies have in some ways benefitted from the wholesale destruction of natural entities, restoration by such individuals is still best seen as an attempt to make amends rather than bestowing gratuitous benefits.

4   Restoration Presupposes a Destructive Relation

Finally, restoration's positive vision for the human/nature relation fails as it rests on a prior destructive relationship with nature. As we have seen, Jordan argues that restoration is an avenue for human "re-entry" into nature, that restoration is a paradigm for a "healthy" human relationship to nature, that it is a model for human "community" with nature, and that it is the basis for a "new communion with nature."

Although not advocating restoration as a paradigm for the human relation to nature, Andrew Light (2000) also has characterized restoration as a positive relation to nature. He has promoted restoration for its ability to recapture what he calls "the culture of nature," by which he means (in part) humans being in nature, working with it, and thereby coming to understand more about it. Light distinguishes between malicious and benevolent restorations: Restorations are malicious when they are used to justify past harm to nature, whereas benevolent restorations are "undertaken to remedy a past harm done to nature although not offered as a justification for harming nature" (54). According to Light, "benevolent restorations . . . are valuable because they help us restore our relationship with nature" (67). "Restoration," he says, "is an obligation exercised in the interests of forming a positive community with nature" (ibid.).

Focusing solely on the positive dimensions of restoration ignores its essentially regrettable character. Although restitution and reparation are important parts of healthy communities, they only exist to rectify mistakes made by community members. An ideal community would not need such institutions. We certainly would not build healthy relationships and community with others based on policies that promoted members harming each other so that restitution becomes possible. The same point applies to our relationship with nature. Restoration involves an attempt to undo harm. Thus the restoration paradigm suggests that the proper role for humans in nature is first to degrade nature, then to attempt to fix it. This is not a positive vision of humanity's role in nature.[24] Humans must find a type of participatory relationship with nature that doesn't presuppose degrading nature to begin with.

Conclusion

What is needed is a conception of human flourishing that does not feed on the wholesale destruction and domination of nature. We must reject the supposition that culture, civilization, and technology—essential aspects of what make us human—necessarily destroy or dominate nature. If this were true, then perhaps restoration—or pure preservationism's human/nature apartheid—would be the best we could do in our relationship with nature. It is noteworthy that both Jordan and Katz—respectively, the strongest advocate of the restoration paradigm and a major defender of pure preservationism—share this faulty assumption that humanity necessarily degrades and dominates nature.

To understand the possibility of such a positive relationship between humans and nature we need to distinguish between respectful human *use* of nature and human *abuse* of nature. We must distinguish between human influence on, modification of, and involvement with nature on the one hand, and human domination and control of nature on the other. Humans can use nature and be involved with it while respecting its autonomy, as long as they do not massively impact nature or try to dominate or control it. It is only the abusive, domineering human impacts on nature that require restoration.

A key dimension of such a healthy, nondestructive, and nondomineering human relation with nature is ensuring an appropriate human scale on the planet. This would involve many fewer people, modest consumption levels, and use of the most environmentally friendly technologies. In this context, human impacts on nature would not be unfair or otherwise wrong. Far greater levels of wild nature would flourish than do so in today's overpopulated, overconsumptive societies, with their inefficient and lethal technologies. Humans would not be extirpating other forms of life or wiping out ecosystem types. We would not be altering (or managing) global climate or arrogantly spewing our pollutants all over the planet. Humans would certainly be using nature and altering its course, but only on a local scale with harmful impacts limited to individuals.

On such a scale, human use of nature could be respectful and just, and thus it would not require restitution or reparations. Restoration of nature would not be morally required and would seldom be useful.

Nature could typically heal itself from the minimal harms we cause it when we live modestly and responsibly on the planet.

Humans have caused and continue to cause massive damage to nature. Preservationists are right to insist that this arrogant and unjust despoliation must stop. We can and ought to help nature heal from this assault and the restoration movement is a praiseworthy acknowledgment of this power and responsibility. But restoration is a short-term and fundamentally regrettable way of relating to nature. Put forward as an ideal for the positive human relationship to nature, it is grandiose, hubristic, and insensitive to the value of wild nature and to the limited role restoration can play in that relationship. In addition to the fundamental problem that restoration rests on past abusive treatment of nature, global climate change may well limit the possibility of restoration as an appropriate response to this abuse. Ron Sandler, in chapter 3 of this volume, persuasively argues that global warming will make reconciliation an increasingly important environmental virtue. While I think there is clearly an ongoing role (and obligation) for restoration in the short term, putting up with and working with the ecological changes we have caused (Sandler's "reconciliation") will often make more sense than attempting to strong-arm species and ecosystems into climates into which they no longer fit (or geoengineering climates appropriate to them).

I conclude that restoration plays only a minor role in a healthy human/ nature relation. Restoration as an ideal of the paradigmatic relationship with nature only makes sense given the current and past abusive human treatment of nature. Virtuous human flourishing on the planet would not include restoration of nature as a central feature. While much can be learned from the movement to restore nature—particularly how to avoid the pitfalls of pure preservationism—restoration does not provide a paradigm for the ideal human relationship with nature.

Notes

1. I focus on Jordan's account of restoration in 1986, 1994, 2000, 2002, and 2003.

2. Elliot has moderated his anti-restoration views significantly in his more recent book (Elliot 1997), arguing that a restored nature lacks a particular (though very important) component of natural value (namely, a nonhuman genesis) and not that it lacks all natural value.

3. I focus on Katz's views of restoration in [1992] 1997, 2000, 2002, and 2007.

4. There have been many critiques of Katz's views on restoration. For one of the most useful, see Vogel 2003.

5. I do not intend to embrace the idea that nature has a good of its own in anything like the sense in which sentient animals do. Acting for the sake of nature or to benefit nature should be interpreted as acting in a way that increases nature's value in a way other than its instrumental value to humans.

6. I am also skeptical of Thompson's idea that it is desirable for humans to take responsibility and manage the global climate. Even if it were true that "collectively our actions determine the basic conditions for the existence of all life on Earth" (Thompson, chapter 10, this volume), a claim that seemingly ignores the massive contribution of nature to these conditions, it would be far more desirable (and virtuous) if humans were to limit their use of this Promethean power and do what they can to return control of these conditions to natural processes. I view humans taking responsibility for the global climate as a vice: it is an excess of responsibility, a hyperresponsibility, that contributes to undermining the centrally important, given, gifted character of our world. Compare Gardiner, chapter 12, this volume.

7. See Ross 2006 for some useful analysis and examples of this idea.

8. For a discussion of some of these affects, see both Sandler, chapter 3, and Bendik-Keymer, chapter 13, both this volume.

9. For discussion and defense of the importance of historical fidelity in restoration, see Throop, chapter 2, this volume.

10. I do not think Elliot intended his remark about how we should relate to restored ecosystems to be generalized to our relation to all of nature on earth.

11. Kane is criticizing the view expressed in the quote, not embracing it.

12. Eric Katz argues that it is dangerous to articulate a positive vision of humans' role in nature (2002, 143). Given the extent of human manipulation of the earth and the extensive damage we have caused, there is risk in promoting a positive vision of humans' place in nature. It is imperative that humans reduce their impact on the natural world and that we clean up our mess. But humans live on earth and must use nature to survive. We need a theory that characterizes an appropriate human *use* of nature. Katz's art appreciation model and, more generally, the preservationists' visitor-in-a-wilderness model offer no help in this regard. (But Kawall, chapter 11, this volume, might.—eds.)

13. Alternatively, preservationism might embrace a "high-tech" model and advocate the use of the most sophisticated modern technology for the purpose of limiting human impact and involvement with nature.

14. This criticism is only fairly aimed at the primitivist (i.e., low technology) version of preservationism. As noted above, preservationists could embrace and use science and technology to limit our interaction with nature, as well as to study, understand, and appreciate it.

15. Jordan is here quoting Eiseley 1970, chap. 7.

16. However, many native societies did not avoid the tragedy of the commons, a point that leads one to question human foresight. Compare Hirsch and Norton, chapter 2, this volume.

17. For a powerful critique of the methods and goals of one restoration project with which Packard was involved, see Mendelson, Aultz, and Mendelson 1992.

18. For a defense of understanding restoration in terms of a healing metaphor, see Throop, chapter 2, this volume.

19. Paul Taylor (1986, 114–115) makes this case powerfully:

It seems quite clear that in the contemporary world the extinction of the species *Homo sapiens* would be beneficial to the Earth's Community of Life. The destruction of natural habitats by housing developments, industrial complexes, airports, and other large-scale projects would cease. The poisoning of soil and pollution of rivers would come to an end. The Earth would no longer have to suffer ecological destruction and widespread environmental degradation due to modern technology, uncontrolled population growth, and wasteful consumption. After the disappearance of the human species, life communities in natural ecosystems would gradually be restored to their former healthy state. Tropical forests, for example, would again be able to make their full contribution to a life-sustaining atmosphere for the whole planet. The lakes, oceans, and wetlands of the world would slowly become clean again. Spilled oil, plastic trash, and even radioactive waste, after many centuries might finally cease doing their terrible work. . . . Our presence, in short, is not needed.

20. It is worth considering to what extent the preceding criticisms of the restoration paradigm's suggestion that nature as a whole is dependent on us to take care of it applies to Allen Thompson's idea (chapter 10, this volume) that human goodness in today's world requires that we responsibly manage the earth's global climate and basic ecological conditions.

21. For a defense of the value of wildness/naturalness, see Hettinger and Throop, 1999. For a discussion of the importance of the distinction between human influence on nature (its degree of naturalness) and human control of nature (its lack of autonomy), see Hettinger 2005. Steven Vogel (chapter 15, this volume) raises a host of provocative objections to the distinction between humans and nature that the value of wildness presupposes.

22. Jordan abandons the naive garden metaphor and directly addresses the problem that gardening suggests an unacceptable type of control and manipulation of nature: "Restoration is not . . . domestication. It does . . . involve manipulation and is a form of agriculture . . . but . . . it is agriculture in reverse. If the gardener . . . takes charge of the landscape, the restorationist does just the opposite. . . . Restoration amounts to a deliberate attempt to liberate the landscape from management" (Jordan 2000, 27–28). He therefore helpfully characterizes restoration as "re-wilding" rather than gardening. In his later book (Jordan 2003), Jordan distinguishes restoration from traditional gardening (which he argues is a creative activity) by arguing that restoration "is valuable as a special form of gardening that is . . . explicitly *noncreative* with respect to objectives, neither improving on nature nor improvising on it but attempting, blankly, to copy it" (24). He also reiterates his idea that restoration attempts to free nature from human control: "While agriculture ordinarily involves bringing nature under control to a certain extent, simplifying an ecosystem in order to exploit

it more effectively for some human end, restoration does just the opposite, recomplicating the system in order to set it free, to turn it back into or over to itself, with a studied indifference to human interests" (87). For an evaluative survey of guiding metaphors in restoration ecology, see Throop, chapter 2, this volume.

23. The notion that restoration is a gift back to nature continues throughout Jordan 2003.

24. While benevolent restorations (in Light's sense) should often be undertaken and are the right thing to do, they are attempts to make the best of a bad situation. Light uses the analogy of a human carrier of disease who ignores warnings and infects other people. He then asks: "Would it not in the end benefit her to volunteer in a hospital ward full of people dying from this particular disease?" (2000, 66–67). While the answer is clearly yes and (as Light points out) she would learn a lot about the wrong she had done, this is not a model for positive community relationships. Restoration is a necessary step as we begin to repair our relationship with nature. But a future positive human/nature relationship would avoid the need for restoration in the first place.

References

Diamond, Jared. 1992. Must We Shoot Deer to Save Nature? *Natural History* 101 (8): 2–6.

Eiseley, Loren. 1970. *The Invisible Pyramid*. New York: Charles Scribner's Sons.

Elliot, Robert. 1982. Faking Nature. *Inquiry* 25: 81–93.

Elliot, Robert. 1997. *Faking Nature: The Ethics of Environmental Restoration*. London: Routledge.

Gould, Stephen Jay. 2001. The Golden Rule: A Proper Scale for Our Environmental Crisis. In *Environmental Ethics: Readings in the Theory and Application*, ed. Louis Pojman, 214–218. Stamford, CT: Wadsworth. Originally appeared in *Natural History* (September 1990).

Hettinger, Ned. 2005. Respecting Nature's Autonomy in Relationship with Humanity. In *Recognizing the Autonomy of Nature*, ed. Thomas Heyd, 86–98. New York: Columbia University Press.

Hettinger, Ned, and Bill Throop. 1999. Refocusing Ecocentrism: De-emphasizing Stability and Defending Wildness. *Environmental Ethics* 21: 3–21.

Jordan, William R. III. 1986. Restoration and the Reentry of Nature. *Orion Nature Quarterly* 5 (2): 14–25.

Jordan, William R. III. 1992. Editorial: Otro Mundo; Restoration, Columbus, and the Search for Eden. *Restoration and Management Notes* 10 (1): 3.

Jordan, William R. III. 1994. Sunflower Forest: Ecological Restoration a Basis for a New Environmental Paradigm. In *Beyond Preservation: Restoring and Inventing Landscapes*, ed. A. Dwight Baldwin, Judith de Luce, and Carl Pletsch, 17–34. Minneapolis: University of Minnesota Press.

Jordan, William R. III. 2000. Restoration, Community, and Wilderness. In *Restoring Nature: Perspectives from the Humanities and Social Sciences*, ed. Paul Gobster and Bruce Hull, 23–36. Washington, DC: Island Press.

Jordan, William R. III. 2002. Restoration as a Responsibility: An Interview with Bill Jordan III. *Orion Afield* (Spring): 25–27.

Jordan, William R. III. 2003. *The Sunflower Forest: Ecological Restoration and the New Communion with Nature*. Berkeley: University of California Press.

Kane, Stanley. 1994. Restoration or Preservation? Reflections on a Clash of Environmental Philosophies. In *Beyond Preservation: Restoring and Inventing Landscapes*, eds. A. Dwight Baldwin, Judith de Luce, and Carl Pletsch, 69–84. Minneapolis: University of Minnesota Press.

Katz, Eric. [1992] 1997. The Big Lie. In *Nature as Subject*, 93–107. Lanham, MD: Rowman and Littlefield.

Katz, Eric. 2000. Another Look at Restoration: Technology and Artificial Life. In *Restoring Nature: Perspectives from the Humanities and Social Sciences*, ed. Paul Gobster and Bruce Hull, 37–48. Washington, DC: Island Press.

Katz, Eric. 2002. Understanding Moral Limits in the Duality of Artifacts and Nature: A Reply to Critics. *Ethics and the Environment* 7 (1): 138–146.

Katz, Eric. 2007. Book Review: *The Sunflower Forest: Ecological Restoration and the New Communion with Nature* by Will R. Jordan. *Ethics and the Environment* 12 (1): 97–104.

Light, Andrew. 2000. Ecological Restoration and the Culture of Nature: A Pragmatic Perspective. In *Restoring Nature: Perspectives from the Humanities and Social Sciences*, ed. Paul Gobster and Bruce Hull, 49–70. Washington, DC: Island Press.

McKibben, Bill. 1989. *The End of Nature*. New York: Doubleday.

McKibben, Bill. 1995. *Hope, Human and Wild: True Stories of Living Lightly on the Earth*. New York: Little, Brown, and Company.

Mendelson, Jon, Stephen Aultz, and Judith Mendelson. 1992. Carving Up the Woods: Savanna Restoration in Northeastern Illinois. Reprinted in *Environmental Restoration: Ethics, Theory, and Practice*, ed. Bill Throop, 135–144. Amherst, NY: Prometheus Books.

Nabhan, Gary Paul. 1991. Restoring and Re-storying the Landscape. *Restoration and Management Notes* 9 (1): 3–4.

Packard, Steve. 1990. Guest Editorial: No End to Nature. *Restoration and Management Notes* 8 (2): 72.

Packard, Steve. 1993. Restoring Oak Ecosystems. *Restoration and Management Notes* 11 (1): 5–16.

Pollan, Michael. 2003. Jacket endorsement. On *The Sunflower Forest: Ecological Restoration and the New Communion with Nature* by William R. Jordan III. Berkeley: University of California Press.

Rolston, Holmes, III. 1994. *Conserving Natural Values*. New York: Columbia University Press.

Ross, Stephanie. 2006. "Paradoxes and Puzzles: Appreciating Gardens and Urban Nature." *Contemporary Aesthetics* 4. http://www.contempaesthetics.org/newvolume/pages/article.php?articleID=400 (accessed 8/26/2011).

Taylor, Paul. 1986. *Respect for Nature*. Princeton: Princeton University Press.

U.S. Congress. 1964. The Wilderness Act.

Visvader, John. 1996. Natura Naturans: Remarks on the Nature of the Natural. *Human Ecology Review* 3 (Autumn): 16–18.

Vogel, Steven. 2003. The Nature of Artifacts. *Environmental Ethics* 25 (2): 149–168.

2

Environmental Virtues and the Aims of Restoration

William M. Throop

The practice of ecological restoration provides an important arena within which we can work out what it means to have a moral relationship with nature today (see Throop 2000; Hettinger, chapter 1, this volume). The practice involves responding to damage for which we are responsible; it involves morally significant relations to individuals and groups—human and nonhuman—and it engages us in deliberation about who we should be in relation to evolving systems. How and when we restore reflects our images of nature and of human flourishing. Yet climate change complicates our attempts to develop a defensible moral relationship with nature, because it threatens to undermine both traditional views about how nature fits into a good human life and traditional aims of restoration. I will focus primarily on the latter threat, although my argument has implications for the former.

In some ecosystems, the threat of climate change raises doubts about the wisdom of using predisturbance conditions to fix restoration goals, since native communities may not flourish in a shifting climate (Harris et al. 2006). The 2007 IPCC assessment report projects a low estimate for warming in the next century of 3.2°F, and a high estimate of 7.2°F. Even at the low estimate, we will see significant shifts of species, and 20–30 percent of species will be at increased risk of extinction.[1] With significant uncertainty about the viability of historical ecosystems, restorationists need to clarify and defend their traditional historical constraints on the selection of project goals or reject them.

For instance, in The Nature Conservancy's (TNC) large clayplain forest restoration project in central Vermont, the goal has been to return the exact species mix located in the remnant forests that dot the ancient lake sediments east of Lake Champlain. TNC has created a nursery to grow native plants from seed sources in the area to assure that as nearly as possible historical conditions are replicated. Sugar maples are an occasional species in clayplain forests, and they are an important part of the lives of Vermonters. Since seed sources have been limited, sugar maples have constituted a relatively small number of the trees planted in recent restoration projects. If Vermont's climate warms significantly during this century, however, sugar maples will not likely be viable in the clayplain forest.[2] Under these circumstances, should restorationists plant sugar maples at all? If so, should they come from native stock?

If we insist on returning to a predisturbance system, we may increase the failure rate of restorations and threaten funding for large projects. Climate change may actually liberate ecosystem managers from the past and enable them to create novel ecosystems that maximize utility. Indeed, in a globalized world characterized by massive movements of goods, peoples, and organisms, perhaps we should abandon altogether "purist" ideas about the way ecosystems should be, and instead emphasize protection of the ecosystem services they provide for us.[3]

Some restoration leaders have seen climate change as an opportunity to reorient restoration. For example, Hobbs and Harris (2001, 239) argue, "Setting clear and achievable goals is essential, and these should focus on the desired characteristics for the system in the future rather than in relation to what these were in the past."[4] Although they are clearly right about the importance of selecting achievable goals, I will argue that the call for a restoration "about-face"—a turn from the past to the future—is inappropriate and that it would lose a critically important moral justification for restoration. Although climate change should shape the way history constrains goal selection, it should not unhinge restoration from history.

My central line of reasoning will proceed as follows:

1. I will *assume,* based on other work, that our selection of restoration goals should be based, in part, on moral considerations regarding our relations to nature.

2. Then I will *argue* that a healing metaphor for restoration seems to best express those moral dimensions of restoration associated with our relations to ecosystems.

3. I will also argue that the main virtues associated with the healing metaphor are humility, self-restraint, sensitivity, and respect for the other.

4. Finally, I will conclude that if we select restoration goals in ways that express these virtues, then we will place a premium on historical fidelity as a criterion of good restoration.

Before launching into this argument, I must clarify the sense in which history is important to restoration. Eric Higgs (2003) is an excellent guide to the shifting conceptions of the role of history in ecological restoration.[5]

Predisturbance Conditions and Historical Fidelity

In 1990, the Society for Ecological Restoration (SER) adopted a definition of restoration that focused on the historically accurate end product of a restoration process—a recreated indigenous ecosystem. "Ecological restoration is the process of intentionally altering a site to establish a defined indigenous, historic ecosystem. The goal of the process is to emulate the structure, function, diversity and dynamics of the specified ecosystem" (Higgs 2003, 107; see also Throop 1997 and Hettinger and Throop 1999). Higgs shows how this restrictive view of restoration fails to encompass the variety of contexts in which restoration occurs and the range of meanings that it can have. In urban settings and in highly humanized rural landscapes like much of Europe, returning to an indigenous ecosystem often makes little sense. Still, land managers should act to ameliorate ecological damage.

The 2002 SER definition provides a much more inclusive account of ecological restoration that enables restorationists to retain human features of the landscape and that acknowledges the dynamic nature of ecosystems: "Ecological restoration is the process of assisting the recovery of an ecosystem that has been degraded, damaged or destroyed."[6] This definition increases the creativity that restorationists can exhibit in the face of ecological damage, broadens the appeal of restoration projects by permitting stakeholders more input into goal setting, and enables projects to be more responsive to the constraints imposed by the context. It does not mention historical constraints, however, although these may be implicit in the notion of recovery.

Higgs's own account makes these constraints explicit. Good restoration must exhibit historical fidelity, that is, "loyalty to predisturbance conditions, which may or may not involve exact reproduction" (Higgs 2003, 127). There are many ways of being "loyal" that diverge from past ecosystem structures, and the constraints on what counts as loyalty are largely contextual. This permissive account of historical fidelity reflects Higgs's vision of continuous landscape evolution resulting from the interplay of cultural and ecological factors. According to this vision, restorationists should seek to move an ecosystem forward toward a structure that reflects what people value about the past, while responding to changing situations. Managers should not try to move an ecosystem backward toward some idealized "natural" structure. Instead, managers should engage stakeholders in planning, designing, and executing a restoration project, using public participation as a dominant mechanism for infusing cultural elements into the recovery of damaged ecosystems.

Such a vision of the role of history in restoration permits selecting goals that accord with projected climate changes. It sets a general direction for restoration, but it should be more specific if it is to guide goal selection. I suggest that the following more specific criteria for goal setting amid climate change accord with Higgs's historical fidelity.

One is faithful to the historic ecosystem if one selects as a restoration goal an ecosystem that (1) is viable under the probable future climate and other contextual changes anticipated to occur in the intermediate term (e.g., fifty years); (2) would likely have evolved under gradual climate change in the absence of other forms of degradation; and (3) is as close as possible to the predisturbance system and a reasonably close approximation to it. In the first clause, viability would be determined by criteria such as resilience, elasticity, and stress response (cf. Higgs 2003); the restoration goal must yield a system that functions appropriately within a time frame suitable for measuring success of the project. The second clause constrains goals according to ecosystem types that would likely evolve in the region as a result of more gradual climate change, such as through migration of species. The third clause appeals to an intuitive notion of closeness to the predisturbance system, which may be measured along a number of dimensions, including the proportion of species/genotypes shared by goal and predisturbance systems, the similarity of species that are not shared (similarity in terms of functional role and evolutionary heritage), and the geographical proximity of other instances of the goal system to the predisturbance system.

The three clauses are in order of priority, with the first two setting constraints within which we should seek to approximate historical ecosystems. So understood, historical fidelity is still just one of several criteria for goal selection. So a project may count as good restoration even if it has limited historical fidelity. I shall argue, though, that from a moral perspective it is a particularly important criterion.

Higgs defends the historicity condition on ecological and social grounds. Recovery of prior ecosystem structures may be the best way to promote ecological integrity, though climate change may affect the likelihood that this is true. Social values of nostalgia, narrative continuity, and awareness of "deep time" also play a central role in justifying loyalty to past structures.[7] These values are often best expressed by ecosystem structures that are continuous with past ecosystems but that reflect current ecological and social realities.

Higgs describes nostalgia as "a bittersweet longing for something lost." The *Vermont Life* image of my state seems an apt example and has indeed justified the protection of historical ecosystems. Narrative continuity comes from the stories we tell about a place that express its significance for us and that tie us to it. Time depth refers to our deepening affiliation with a person, place, or object that comes with a long relationship. We are often moved by such considerations, but since they rest in human interests, they may easily be trumped by other human interests that would violate historical fidelity. For example, because of time depth and nostalgia, my closet has too many pairs of old hiking boots that I no longer use. If I needed the space, however, those boots would be gone; time depth would be easily trumped. Boots are not ecosystems, and our relations to the land can be much more powerful. Still the moral dimensions of those relations are less likely to be trumped by our other interests, so they ought to carry much of the weight in our decision making.

Ecological restoration seems like more than just "a good thing to do"; it feels like something we *ought to do* when we have seriously degraded an ecosystem. No doubt part of this sense of duty comes from obligations to humans who depend on the land, but part also seems to come from our sense of who we ought to be *in relation to the land*. This accords with the intuition that the burden of proof would lie on those who believe they could degrade an ecosystem and just walk away without moral consequence, as long as no humans are hurt. If we do have a duty to restore, then moral considerations ought to play a significant role in the goal selection for restoration.

Healing as a Guiding Moral Metaphor for Restoration

Virtue ethics provides a useful lens through which to view the moral dimensions of goal setting, because it highlights many of the morally significant elements of our relations to nature. But what virtues should restorationists have? Should courage or justice, creativity or fidelity, boldness or humility, guide our restoration planning? Often a central metaphor that guides a practice highlights its dominant virtues. The metaphor of a competitive game characterizes the practice of much law in the United States. Thus the virtues associated with effective treatment of adversaries dominate the practice. With ecological restoration, by contrast, three quite different metaphors have been advanced to define its practice: gardening, design, and healing. Each emphasizes different virtues, and each is suited to some contexts. Overall, however, the healing metaphor seems to capture best the primary moral motivations for restoration, as well as current ecological realities.

The gardening image figures prominently in the thinking of Bill Jordan (1994; 2003) and Fredrick Turner (1985), and it is behind the scenes in many of the articles in Gobster and Hull (2000). Jordan characterizes his view as follows:

> Briefly, what Fred [Turner] was suggesting was that the act of gardening offers a model for a healthy relationship between human beings and the rest of nature. His argument, in part, was that the gardener handles nature with respect but without self-abnegation—that is, he or she manipulates nature intelligently and creatively, benefiting and nurturing plants . . . while at the same time exercising a wide range of human aptitudes and leaving a distinctively human mark on the landscape. . . . By the time I read Fred's essay, I had already identified restoration as a form of gardening. (1994, 17)

The gardening metaphor suggests that states of nature have value only because we value them (Throop 2001). We value ecosystems for many different reasons, but all the relevant values in restoration are values for us. There is no single correct goal for restoration as gardening, and since goals are negotiated, ideally restoration should involve the participation of stakeholders in all phases of the project, including goal setting (Light 2000). Part of the value in gardening is the activity. Since on this image restoration aims to build a healthy relation between humans and nature, it should involve lots of humans.

The virtues of a gardener include knowing the range of goals that could be achieved and how to achieve them, being able to effectively integrate people into the restoration process, and being able to listen to people and to lead them. One model solution might be the Montrose Point case in Chicago (Gobster and Barro 2000). There, stakeholders were included in the project from the beginning. They forged a compromise about restoration goals that protected their various interests and thereby avoided the gridlock that occurred in the Chicago Wilderness project.

The main problem with the gardening metaphor, however, is that it is highly anthropocentric. It does emphasize some moral dimensions of our relation to nature, but it tends to elevate ecological services that nature provides for humans, and not the values associated with nonhuman nature. It is particularly well suited for urban restoration, where human interests are bound to dominate. But in much restoration, where the focus is on the ecosystem itself, more human self-abnegation is probably a good thing.

The design metaphor is implicit in much of the ecological engineering literature (cf. Mitsch and Jorgensen 2004), and explicit in Higgs (2003). It too emphasizes the human role in restoration and the creativity it involves. We design restoration projects to satisfy a range of needs and values within specified constraints. Higgs emphasizes design, because he thinks that it serves as an antidote to the self-abnegation that character-izes some thinking about restoration: "We need to acknowledge that restoration is fundamentally a design practice. Abnegation is not the proper path—we should celebrate and enlarge the skills and wisdom in restoration design, not bury it under a patina of ecological accuracy" (Higgs 2003, 274).

Where some people want to minimize the human element in restora-tion, Higgs wants to dramatize the fact that in restoration, we humans "inscribe our intelligence on the landscape" (2003, 270). Higgs defines "design" as an intentional activity that is carried out in a skillful or artistic way. Few would deny that restoration involves design in this sense. But Higgs emphasizes a stronger sense of "design" in which a restored ecosystem is itself crafted with a specific result in mind, and the process of restoration is designed so that it includes volunteer participation.

The virtues of a designer include the disposition to comprehend and control the whole system being designed and to carefully craft the inter-relationships between its parts. The problem with the metaphor, though, is that it suggests too much knowledge and control, too much precision, and like gardening, too much emphasis on the human. Ecosystems are complex systems governed to a considerable degree by nonlinear dynam-ics (Botkin 1990). As a result, we are often not aware of the long-term impacts of an ecosystem manipulation that alters its structure or com-position. We also have limited ability to control an ecosystem once a new process is introduced.[8]

The metaphor of design seems an inapt way to capture the human intentions, skills, and creativity that figure in good restoration. Many intentional activities done with skill and creativity do not seem appro-priately characterized as design in the strong sense. For example, in jazz improvisation, a musician may exhibit considerable skill on an instru-ment, and creativity in her ability to respond and improvise within a musical framework. Still, it seems counterintuitive to say that a jazz improvisation is designed, in part because it involves too much spontaneity and responsiveness to other musicians. Similarly in trauma surgery, a doctor may have a plan, but people do not typically speak of *designing* the surgery, for design connotes more control and imposition than is usual in such cases.[9] Much restoration seems more like jazz or trauma surgery than design.

By contrast, the metaphor of healing, which dominated the early restoration literature, emphasizes respect and care for the other, responsiveness to its situation, and limited control over the results. Stephanie Mills describes salmon restoration in the Pacific Northwest in the following way:

Now the headwaters, says David Simpson, a founder of the Mattole Watershed Salmon Support Group, is in healing mode. . . . "All we're doing is trying to hang on to what we've got while the habitat restores itself. It's the only hope of recovery." Simpson likened the nature of this change to the long, slow turn of a battleship. "All of human history is directed toward ecological disrepair," he said, so it is hardly surprising, although no more bearable, that redirecting our lifeway toward healing should be starting when it's nearly too late. (1995, 166–167)

This metaphor also figures centrally in Richard Nilsen's (1991) *Helping Nature Heal* and in Holmes Rolston's (1994) *Conserving Natural Value*, where ecological restoration is likened to setting a broken bone. Simpson's observations capture important aspects of the healing metaphor. Nature does much of the restoration work, and humans are just helping create the circumstances where nature can heal. It also suggests that restoration goals should be highly constrained by the nature of an ecosystem; so one's choice of goals is largely determined by one's view about what is "good for the ecosystem," just as the physician aims at the good of the patient.

Where in gardening the emphasis was on the relationship between humans and nature, here it is on the ecosystem itself. In this way, the healing metaphor seems to capture the sense of duty we have to repair our relation to nature when we have degraded it. Also gardening and design do not typically need a specific impetus. They may be acts of creativity, but health is typically disturbed by some specific event and healing returns the system to its functioning prior to the disturbance. There is a focus on addressing a specific problem—like a wound—rather than a range of desires. Since restoration is occasioned by some serious human-caused ecosystem disturbance, it is better characterized by the healing metaphor.

The healing metaphor has its own limits. If it is taken too literally, it suggests that ecosystems are like organisms, which contravenes much of contemporary ecology (Suter 1993; Wicklum and Davies 1995). *I take it to be a moral metaphor, not an empirical metaphor.* It can guide our moral approach toward ecosystems without guiding our empirical research. Thus I am not implying that there is a specific healthy state of an ecosystem. I am also not aligning restoration with standard practices of modern medicine, much of which takes a mechanical approach to treating patients and exhibits a hubristic faith in technology. The healing metaphor is more clearly expressed in alternative approaches to medicine and in the attitudes of many old-fashioned general practitioners.

Healing Virtues

So understood, the healing metaphor embodies a number of related virtues: humility, respect for the autonomy of the "patient" and a corresponding tendency to restrain self-interest, and sensitivity to the idiosyncrasies of a system and its surroundings. It is misleading to say that these virtues are derived from the metaphor. They become salient as we unpack the implications of metaphor as it applies in the context of ecological restoration.

Humility, as I understand it, is primarily an epistemic virtue. It involves the recognition of one's limits, whereas self-restraint enjoins limiting the pursuit of one's desires. Although humility is typically associated with recognition of limited self-importance, the limits to knowledge and to the ability to control one's surroundings are equally relevant.[10] A humble person is regularly aware of such limits and acts accordingly. The braggart lacks humility, but so too does the outwardly modest person who believes falsely that she knows much more than she does or who presumes to have abilities that far outstrip her talents. Indeed the latter flaws are most closely identified with the hubris that carries disaster in its wake. When we fail to recognize the variety of ways in which we are limited, our presumptuousness often results in failure. This is the primary reason we value humility in a healer. Healers proceed carefully and avoid presuming that they have the power to fix a problem. Given the limitations of our ecological knowledge, especially under conditions of climate change, humility seems an important virtue for the restorationist.

Second, by definition, healers aim at the good of another; thus they must exhibit the virtue of self-restraint. The dispositions to recognize when we have had enough and to stop when inclined to go forward are central to the virtue of self-restraint.[11] As our power to satisfy our desires increases, we must be more vigilant regarding self-restraint, for we cannot count on the fear of frustrated desires to motivate us. In a society devoted to consumption, the need for such vigilance is further increased, though its recognition is often belated. In restoration it is easy to let our own interests dominate the design of a project, for we must negotiate goals with many stakeholders. Humans have a tendency to overreach, to grasp for more than they need or can handle. The healing metaphor erects barriers to such overreaching, because it takes respect for the other—the patient—as a guiding principle.

Third, a good healer exhibits sensitivity to the particularities of the patient. Overtly similar systems may react differently to interventions so a healer must listen carefully to the patient, tailor a response, and be ready to adapt to unexpected reactions. Thus the healer cannot adopt a mechanical approach to problem solving, identifying the symptoms of a disease and adopting a standard response. This is especially important when the "patient" is an ecosystem, interrelated with other systems in the region. Such sensitivity is a natural outgrowth of a humility regarding ecological understanding and restraint on our self-interest.

The healing metaphor for restoration, as expressed by these three virtues, coheres well with the state of our empirical knowledge and with the acknowledgment of the moral dimensions of our relationship with ecosystems. It seems most apt for relatively wild ecosystems (Throop 2001; Throop and Purdom 2006), for these are places where the interests of humans should be least visible. It is tempting to say that we should use different metaphors in different restoration contexts, and hence be guided by different virtues. But the healing metaphor seems to capture a primary moral motivation for all restoration projects: *undoing a harm we have caused*. This gives the metaphor a moral primacy in restoration practice. It provides a moral orientation toward restoration, but where our knowledge is considerable or our interests legitimately dominate a project, the other metaphors I have discussed may fruitfully guide our actions. Just as a healer may be guided by design practices in reconstructive surgery, a restorationist may heal some urban park by designing a system that satisfies enough human interests to be sustainable.

Healing Virtues and History

If restorationists act in accordance with the virtues of humility, sensitivity, and self-restraint, they will tend to adopt conservative restoration goals—

those that exhibit a high degree of historical fidelity. And this is the point I have been driving at since the beginning of this chapter. One way to illustrate the connection between these virtues and the criterion of historical fidelity is to see why the abandonment of history courts the corresponding vices. Take humility, for example. The goal of engineering some new, better ecosystem in response to prior degradation requires that we have the knowledge and ability to produce the desired result. Our history of large-scale land manipulation suggests that in many contexts this belief would express hubris. While ecologists often do know the local effects of small-scale manipulations on ecosystems, they typically have limited knowledge of the effects of larger-scale manipulations, especially when the external context is shifting, as with climate change. They may know that certain ecosystems are unlikely to be viable, but they can rarely know that some new system they create will function as they predict. Insofar as ecosystems exhibit nonlinear dynamics, this problem becomes even more acute.

A safer course would be to use one's knowledge to avoid nonviable systems and to reproduce the conditions that permitted the historical system to flourish to the extent that it is possible—in other words, to reset the clock. One could then let the system evolve in the new context, removing barriers to its shifting gradually in response to changing circumstances. This approach seems to presume the least amount of knowledge of how the system functions and only the ability to return it toward it predisturbance condition.

If we abandon the criterion of historical fidelity we also court a version of gluttony—an overemphasis on satisfying our own desires. Once we decide to respond to degradation by creating a new system, human interests are likely to play a large role in shaping the result, since these will be much more salient that the largely silent interests of nonhumans. Such interests often focus on a small subset of the system and appear to be satisfiable without protecting the integrity of the whole system. As a result, the restoration project is less likely to protect species unrelated to human interests and to exhibit respect for the degraded system. This is one of the dangers of the ecological services approach to justifying restoration. But if we practice self-restraint, we will be more likely to be sensitive to the dynamics of the system that are unrelated to our immediate interests and more likely to try to replicate predisturbance dynamics as a means of maximizing its autonomy.

Ronald Sandler (chapter 3, this volume) argues that under conditions of climate change, the virtues that are traditionally associated with ecological restoration, including humility and care for the ecosystem, should lead us to deemphasize historical fidelity. In some cases, an appropriate humility should lead us to recognize our inability to restore viable historical ecosystems as climate changes, and thus we should assist systems in moving toward viable novel forms. There is no question that humility might be manifested in promoting a range of different restoration goals in a large-scale project (see, for example, Cabin 2007). Self-restraint can lead one to minimally manipulate the current system rather than try to return it toward a predisturbance system; sensitivity to the system can suggest that some novel evolution would be less disruptive.

The soundness of Sandler's argument depends in part on empirical questions regarding how fast our climate will change and for what proportion of human-damaged ecosystems, a historically faithful restoration will not be viable within the intermediate term. Since I am not defending exact reproduction of the predisturbance, but rather removing a disturbance (resetting the clock) and letting the system evolve, it is far from clear that climate change is, or will soon be, radical enough to significantly diminish the importance of historical fidelity so understood.

In cases where we do not have positive knowledge that pursuing historical fidelity is an irrational aim, the healing metaphor supports historical fidelity. This presumption in favor of historical fidelity results from viewing restoration as an expression of our moral relationships with ecosystems. Sandler's argument ill accords with an emphasis on the moral dimensions of that relationship; creating a novel system in response to damage suggests that systems are morally interchangeable. But to be in a relation with a system is to value *that* system. Ecosystems are defined in part by their history, as are many other collective entities.[12] If we have damaged a historically defined system with which we have a morally significant relation, then—other things being equal—we should heal *that* system. This is the context within which we express humility and self-restraint.

The force of this argument may explain why so many feel that we should "restore" New Orleans after the damage inflicted by hurricane Katrina. Many of us feel the pull of a duty to rebuild the city rather than moving it or resettling the population, because the city is historically defined and we have a relation with that historically defined entity. This is not just an appreciation of time depth or nostalgia; it has a moral dimension. That moral dimension is not exhausted by our duties to the inhabitants of New Orleans, who wish to return to their city but who might be served through resettlement. Perhaps the presumption in favor of restoring New Orleans should have been overridden by concerns about the long-term viability of the city. But in such a case, it would be a mistake to say that there was no moral reason for the presumption in favor of historical fidelity.

A conservative approach to restoration in cases of climate change will focus on removing that which harmed the system and letting it heal. Most historical ecosystems have withstood the test of time, responding effectively to shifting inputs. The healing metaphor suggests that we should work with that resilience. Climate change may send a system onto a new trajectory, and we may be able to accurately predict many local changes that will occur, such as migration of species. The aim of historical fidelity permits restorationists to take such predicable changes into account, but not to engage in wholesale reconstruction of a system.

Conclusion

Let us return to the question about replanting maple in Vermont's clay-plain forest restorations. If we take seriously the virtues associated with healing, we will replant sugar maples in roughly the percentages found in remnant clayplain forests. This accords with an acknowledgment that we do not know whether the climate will change sufficiently to eliminate sugar maples, and with our limited knowledge of how removing them would shift the system's adaptation to climate change. Of course, we must recognize that they may not flourish over time, but at least the system will evolve on its own primarily in response to the changing climate and not in response to our changing interests and knowledge.

But because climate change will probably significantly affect the viability of the sugar maples in Vermont within fifty years, we should also take steps to remove barriers to gradual evolutionary change in the system. As a result, we probably should not use only the sugar maple stock that is indigenous to the region. The Nature Conservancy plant nursery should probably increase the amount of genetic variation in the maple seedlings used for the restoration project and include varieties that have adapted to southern end of the sugar maple range (Gunter et al. 2000). In general, the goal of returning to a predisturbance ecosystem is interpreted as involving the use of local seed sources,[13] but under conditions of climate change, this principle may not be advisable. Here the healer recognizes that alleviating the damage must take into consideration the likely future trajectory of the system. Historical fidelity need not involve historical replication. In addition, TNC and other organizations should renew their efforts to create migratory corridors for lowland forest species, which may alter the composition of clayplain forests as climate changes.

In *The Economy of the Earth*, Mark Sagoff (1988) maintains that our character as a people plays a critical role in our selection of environmental goals. I have sketched a virtues approach to restoration goal setting under conditions of climate change that illustrates that role and that promises to help us recapture an aim that originally guided the practice of restoration. The aim of historical fidelity remains reasonable in an age of climate change and serves to ground restoration practice in a defensible moral orientation toward ecosystems, that of the healer.

Notes

1. See also Bendik-Keymer, chapter 13, this volume, on the sixth mass extinction.

2. See the climate change tree atlas at   http://www.nrs.fs.fed.us/atlas/tree/tree _atlas.html (accessed August 28, 2011) for projected shifts in 134 U.S. tree species given five different climate models.

3. The appeal to ecosystem services is used increasingly to justify large-scale projects. For example The Nature Conservancy uses ecosystem services to guide large scale river conservation; see http://www.nature.org/ourinitiatives/habitats/ riverslakes/howwework/river-science-ecosystem-services-approach.xml (accessed August 28, 2011). See also Light, chapter 5, this volume.

4. Some philosophers are reaching similar conclusions; Sandler, chapter 3, this volume, argues that global warming elevates the virtues of openness, accommodation, and reconciliation and diminishes the importance of historical constraints on restoration.

5. Higgs, chapter 4, this volume, extends his analysis in ways that complement my main argument.

6. Society for Ecological Restoration International Science & Policy Working Group, The SER International Primer on Ecological Restoration, October 2004, http://www.ser.org/content/ecological_restoration_primer.asp - 3h (accessed August 28, 2011).

7. It is worth considering how deep-time fidelity relates to the matter of expanded time-scale in adaptive management. See Hirsch and Norton, chapter 16, this volume.

8. Higgs is aware of such challenges. To emphasize the priority of ecosystem values, he coins the phrase "wild design" which focuses on the "largely silent interests of ecosystems" and which recognizes that while human agency frames the projects, natural processes take over.

9. In many forms of elective surgery, from cosmetic surgery to joint replacement, design is an apt metaphor precisely because it is less spontaneous.

10. More work has been done on humility than any other environmental virtue, probably because hubris is thought to be the vice most responsible for our environmental destruction. Modern cultures have experienced tremendous growth of

knowledge and skills, so it is easy to forget limitations. Especially when it comes to manipulating ecosystems at large scales, our limits are quite marked. When our limits are obscured and our actions can have dramatic long-term consequences, we should emphasize the value of humility. Thomas Hill (1983) argues that our appreciation for the virtue of humility explains our discomfort with the destruction of natural systems and with a wholly resource-oriented approach to nature. Geoffrey Frasz (1993) further develops and corrects Hill's treatment of humility. Robert Sparrow (1999) develops a critique of terraforming Mars based on its embodying environmental vices such as hubris. (Compare also to Gardiner, chapter 12, this volume.)

11. Bill McKibben's *Enough* (2003) explores a variety of reasons why we should exhibit self-restraint in the development of technologies such as germ-line genetic engineering. Although he does not call self-restraint a virtue, the book can be read as a sustained defense of this virtue.

12. See Throop and Sanderson 2005 for a defense of the historical identity of ecosystems.

13. See for example, the National Park Service restoration guidelines at http://www.nps.gov/plants/restore/pubs/intronatplant/toc.htm (accessed August 28, 2011).

References

Botkin, D. 1990. *Discordant Harmonies: A New Ecology for the Twenty-First Century*. New York: Oxford University Press.

Cabin, R. J. 2007. Science Driven Restoration: A Square Grid on a Round Earth? *Restoration Ecology* 15: 1–7.

Frasz, G. 1993. Environmental Virtue Ethics: A New Direction for Environmental Ethics. *Environmental Ethics* 15 (3): 259–274.

Gobster, P. H. and Susan Barro. 2000. Negotiating Nature: Making Restoration Happen in an Urban Context. In *Restoring Nature: Perspectives from the Humanities and Social Sciences*, ed. P. H. Gobster and B. R. Hull, 185–208. Washington, DC: Island Press.

Gobster, P. H., and R. B. Hull. 2000. *Restoring Nature: Perspectives from the Humanities and Social Sciences*. Washington, DC: Island Press.

Gunter, L., G. Tuskan, C. Gunderson, and R. Norby. 2000. Genetic Variation and Spatial Structure in Sugar Maple and Implications for Predicted Global Scale Environmental Change. *Global Change Biology* 6: 335–344.

Harris, J. A., R. J. Hobbs, E. Higgs, and J. Aronson. 2006. Ecological Restoration and Global Climate Change. *Restoration Ecology* 17: 170–176.

Hettinger, N., and B. Throop. 1999. Refocusing Ecocentrism: De-emphasizing Stability and Defending Wildness. *Environmental Ethics* 21: 3–21.

Higgs, E. 2003. *Nature by Design*. Cambridge, MA: MIT Press.

Hill, T. 1983. Ideals of Human Excellence and Preserving Natural Environments. *Environmental Ethics* 5: 211–224.

Hobbs, Richard J., and J. A. Harris. 2001. Repairing the Earth's Ecosystems in the New Millennium. *Restoration Ecology* 9 (2): 239–246.

Jordan, W. R. 1994. Sunflower Forest: Ecological Restoration as a Basis for a New Environmental Paradigm. In *Beyond Preservation: Restoring and Inventing Landscapes*, ed. A. D. Baldwin, J. De Luce, and C. Pletsch, 17–34. Minneapolis: University of Minnesota Press.

Jordan, W. R. 2003. *The Sunflower Forest: Ecological Restoration and the New Communion with Nature*. Berkeley: University of California Press.

Light, A. 2000. Restoration, the Value of Participation, and the Risks of Professionalization. In *Restoring Nature: Perspectives from the Humanities and Social Sciences*, ed. P. H. Gobster and B. R. Hull 163–184. Washington, DC: Island Press.

McKibben, B. 2003. *Enough: Staying Human in an Engineered Age*. New York: Henry Holt & Co.

Mills, S. 1995. *In Service of the Wild*. Boston: Beacon Press.

Mitsch, W. J., and S. E. Jorgensen. 2004. *Ecological Engineering and Ecosystem Restoration*. Hoboken, NJ: John Wiley and Sons.

Nilsen, R. 1991. *Helping Nature Heal: An Introduction to Environmental Restoration*. Berkeley: Ten Speed Press.

Rolston, H. 1994. *Conserving Natural Value*. New York: Columbia University Press.

Sagoff, M. 1988. *The Economy of the Earth*. New York: Cambridge University Press.

Sparrow, R. 1999. "The Ethics of Terraforming. *Environmental Ethics* 21 (3): 226–245.

Suter, G. 1993. A Critique of Ecosystem Health Concepts and Indexes. *Environmental Toxicology and Chemistry* 12: 1533–1539.

Throop, W. 1997. The Rationale for Environmental Restoration. In *The Ecological Community*, ed. Roger S. Gottlieb, 39–55. New York: Routledge.

Throop, W. 2000. *Environmental Restoration: Ethics, Theory and Practice*. Amherst, NY: Humanity Books, Prometheus Press.

Throop, W. 2001. Balancing Nature and Humans in Ecological Restoration. *Ecological Research* 19: 215–217.

Throop, W., and R. Purdom. 2006. Wilderness Restoration: The Paradox of Public Participation. *Restoration Ecology* 14: 493–499.

Throop, W., and B. Sanderson. 2005. Autonomy and Agriculture. In *Recognizing the Autonomy of Nature*, ed. T. Heyd, 99–120. New York: Columbia University Press.

Turner, F. 1985. Cultivating the American Garden: Toward a Secular View of Nature. *Harper's* (August): 45–52.

Wicklum, D., and R. W. Davies. 1995. Ecosystem Health and Integrity? *Canadian Journal of Botany* 73: 997–1000.

3

Global Warming and Virtues of Ecological Restoration

Ronald Sandler

Global Warming and Adaptation

In this chapter, I explore the implications of global warming for virtues associated with ecological restoration and assisted recovery. In doing so, I begin from the premise that global warming is now part of the ecological present and future of the planet. Returning to climactic trajectories that would have obtained absent anthropogenic greenhouse gas (GHG) emissions, if possible at all, would require massive and, in my view, ill-advised technological interventions (Hegerl and Solomon 2009; Gardiner 2010 and chapter 12, this volume). Therefore, I take seriously the idea that global warming needs to inform our ecological practices and ethics as an ecological reality, not just as something to be avoided, resisted, feared, and lamented. Throughout this chapter, I will refer to *foregone global warming*. Foregone global warming is the amount of global warming or climactic change that we are already committed to in virtue of obtaining levels of GHGs in the atmosphere plus the most optimistic scenario for future GHG emissions, absent massive technological intervention. How much warming there might be beyond that—that is, warming associated with additional GHG emissions—is of course not foregone.

The relevant features of global warming for the purposes of this chapter, as well as those in virtue of which it is on many people's reckoning the ecological challenge of our time, are increased *rate* of ecological change and increased *uncertainty* (or unpredictability) of ecological change, in comparison to that with which human (particularly agricultural) civilizations are accustomed. Because global warming accelerates the rate of change and exacerbates the information deficit and uncertainties about the ecological future, it makes biological and cultural *adaptation* to ecological changes more difficult, for both us and other species.

As long as there has been climate, there has been climate change—and with it ecological change. As long as there have been systems involving living organisms, adaptation has been needed, and it has occurred. However, the magnitude and rate of climactic and ecological change—and the associated adaptation challenges—have not always been constant. At the core of concerns about global warming is the worry that species, ecological systems, and human societies will not be able to meet the adaptation demands of anthropogenic climate change, and that failure to do so will have high social, economic, and biological costs (IPCC 2007).

The challenge of adaptation that is the product of the rate and uncertainty of ecological change is exacerbated by several factors. These include the *types* of ecological changes expected to be produced, such as increased incidence of extreme weather events and increased climatic variability, as well as the possibility of there being "climactic tipping points" that could result in abrupt changes in climatic and ecological trajectories. In addition, the *range* of possible climatic/ecological futures—for example, with respect to surface air temperatures, sea levels, species' ranges, weather events, precipitation, air velocities, and natural resource availability—within a given time frame is broader. Moreover, the challenge of adaptation is exacerbated by several factors, such as other anthropogenic stressors on ecosystems (e.g., pollution and habitat loss) and the social inertia and myopia encouraged by many social and institutional structures (e.g., neoliberal trade agreements, transnational corporations, twenty-four-hour news media, and election cycles).[1]

The challenge of adaptation is thus a complex social, political, cultural, and ecological challenge. It is also the crux of the social and environmental problems associated with global warming. It is because people, populations of nonhuman species, and ecological and social systems cannot adapt quickly enough that there will be increased rates of species extinctions, instability of ecological systems (or ecosystem reconfiguration), agricultural and natural resource insecurities, exposure to severe weather events, incidence of disease, and ecological refugees (and climate injustice) (IPCC 2007).

In this chapter I argue that, when considered through the lens of the challenge of adaptation, the fact that our ecological future is accelerating away from our ecological past with increasing rapidity, and that it is increasingly unclear where it is going, has several implications for virtues associated with ecological restoration and assisted recovery:

1. Global warming *raises* the salience of virtues related to *openness* and *accommodation* that are commonly associated with restoration and assisted recovery.
2. Global warming *weakens* the justification for *historical fidelity* as a virtue associated with assisted recovery.
3. Global warming *raises* the salience of *reconciliation* as a virtue associated with assisted recovery.
4. Global warming *reduces* the prominence of *restoration* among the types of assisted recovery.

Before elaborating on and trying to motivate these claims I need to clarify what I mean by "ecological restoration" and "assisted recovery," as well as what I take to be the relationship between virtue and ecological restoration.

Restoration

It may be that a precise, exceptionless definition of ecological restoration—one that would distinguish all cases of restoration from all cases of all other types of assisted and unassisted ecological recovery—is not possible. But a functional account—one that captures what is central to ecological restoration and is good enough for present purposes—is possible. The account of ecological restoration that I employ in this chapter is largely that developed by Eric Higgs (2003) in *Nature by Design*. It is a conception of ecological restoration that can be applied in urban environments, agricultural environments, aquatic environments, industrial environments, as well as large wilderness environments. Its three core elements are: ecological integrity, historicity, and design.

The ecological integrity element captures the idea that a restoration improves (or, at least, aims to improve) the condition of an area from an ecological perspective by contaminant remediation, removing dams or impermeable surfaces, or increasing biodiversity, for example. An activity that degrades an area or leaves it no better off than it was prior to the intervention (or does not aim to make it better off) is not a restoration— nor is it any other form of assisted recovery. Whether "integrity," "health," or "functioning" is the superior concept for capturing the ecological quality of a space, and how these are best operationalized or measured in practice—for example, in terms of resilience, self-maintenance, population dynamics, energy flows through the system—is not something I address here. Nothing that follows hinges on it. For present purposes, the crucial point is simply that restoration involves improving an area from an ecological perspective (e.g., removing contaminants, increasing biodiversity, or increasing stability).

It is tempting to formulate the ecological dimension of restoration in value terms—that is, that restoration increases the environmental value of an ecological space. But the value of the space might go up while the ecological condition goes down, since there are environmental values other than ecological integrity (e.g., aesthetic, recreational, and scientific values). When this occurs—that is, when a number of environmental values are amplified in a space even as ecological integrity is diminished— it is not an ecological restoration. Thus, it is necessary to distinguish the ecological integrity of a place from its environmental value overall.

The historicity element of ecological restoration captures the idea that restoration returns something to the way, state, or place that it was previously. In the case of ecological restoration, it is returning something of the ecology that previously obtained in a particular place or region; or else establishing something of the ecology that would have obtained in a particular place or region absent anthropogenic degradation. If an assisted recovery does not "return to the past" in some sense—ranging from replication (i.e., approximating as much as possible some previous state) to incorporating (or amplifying) elements from a previous state (such as native species or abiotic features)—then it is not a restoration. As I am using the term in this chapter, then, "restoration" is a type of assisted recovery that is distinguished by historicity. (And assisted recovery is a type of ecosystem management, along with, for example, conservation and preservation.) *A particular assisted recovery is a restoration to the extent that historicity is incorporated into the project*, in intent, process, or product. Other types of assisted recovery—that is, other activities that involve intervening in or around an ecological space in order to help improve its ecological integrity—include revitalization (which, as I am using the term, does not involve a commitment to historicity) and some cases of reclamation.

The third element of ecological restoration is design. Design is similar to ecological integrity in that it is characteristic not only of ecological restoration, but also of assisted recovery more generally. It is inherent to the "assisted" part of "assisted recovery." All restorations involve decisions about where to restore (if at all), what to aim for, and how to go about accomplishing it—and these decisions are made by people. This is so even when the design of the restoration is to replicate the "design" of nature, since we choose which nature, at what time, and in what respects, in addition to our choosing to pattern the recovery on nature's "design" and not other considerations—for example, aesthetics, economic efficiency, recreation, or rarity.

In this way, all restoration (and all other assisted recovery) has anthropocentric dimensions. It is anthropocentrism in a sense inextricable not just from restoration or assisted recovery, but also from environmental ethics and ethics more broadly. We are the ones responsible for making decisions regarding the considerations that we are responsive to and our forms of responsiveness. There is no other option. Even if we try to abdicate the choice (the design) to nature (or god), we choose the aspects of nature (or the theology) to which we "defer" (Vogel 2006). This sort of anthropocentrism is distinct from anthropocentrism that denies the moral considerability or inherent worth of all nonhuman entities or that denies the intrinsic value of human-independent aspects of nature. After all, even if nonhumans (or human-independent parts of nature) do have inherent worth in the strongest possible sense—that is, independent of their being valued—it is still up to us to recognize that value and respond appropriately to it (i.e., to do what we should). Therefore, that design is inextricable from ecological restoration, and that design is inherently anthropocentric (or, perhaps more appropriately, anthropo*genic*), does not itself imply that restoration is problematic on any theory of environmental values.

Virtue and Restoration

A very general account of the virtues is that they are character traits that dispose a person to respond well to values in the world. More specific theories of virtue are developed by providing substantive accounts of "character traits," "dispose," and "responding well," as well as by providing a value axiology (i.e., an account of what sorts of things are valuable and what sorts of value they have).

The value axiology presumed in this chapter is a pluralistic one that includes one's own flourishing, the flourishing of other people, and human-independent environmental values (e.g., aesthetic values or the worth of nonhuman living things). The conception of responding well to value presumed in this chapter is that it typically involves or is related to protecting and promoting value in the world. Therefore, a character trait is a virtue to the extent that its possession is generally conducive to promoting the good, and a character trait is a vice to the extent that it is generally detrimental to promoting the good. It may be that there are other forms of responding well to value in the world—e.g., appreciation.

However, the primary form of responding well to value is to protect and promote it, and this functions as a constraint on other forms. For example, dispositions to respond to beautiful and wonderful aspects of the natural environment in ways that degrade it—as is often done by recreational scuba divers and off-road vehicle enthusiasts—are not virtues, even if they involve recognition of the aesthetic value of the places.

The conception of character traits and dispositions presumed here is that they are states that decide regarding feelings, actions, and emotions (Aristotle [ca. 350 BCE] 1985, book II, chaps. 5–6). To possess or exhibit a particular character trait is to standardly take certain types of considerations (under certain types of circumstances) as reasons (or motivations) for acting, feeling, or desiring in certain ways (Sandler 2007). For example, a person is compassionate if she standardly takes the (undeserved, unnecessary, and unwanted) suffering of others as a reason (or motivation) for acting to relieve the person's suffering and she has the corresponding affect and desire to do so under certain circumstances—for example, when she is in a position to help. A person who typically is unaffected by and ignores the (undeserved, unnecessary, and unwanted) suffering of others is not compassionate, but indifferent. Thus, compassion and indifference—as well as cruelty—are different character traits regarding the same value (i.e., the well-being of others). They dispose their possessor to respond differently to that value. Because the well-being of others is good, a compassionate person is disposed to respond well to value in the world (i.e., by promoting it), whereas an indifferent person is disposed to respond poorly to value in the world. Compassion is the virtue; indifference is a vice.

Virtue is relevant to restoration in at least three respects. First, some character traits may be virtues in part because they promote value in the context of restoration (or assisted recovery more generally). If ecological integrity is a value and some character traits regarding assisted recovery promote that value better than others, then those character traits are more justified—and more virtuous—than are the alternatives. Second, and related, some character traits make for good restorationists and help to explicate what constitutes good restoration. A good restoration expresses (or hits the target of) the operative virtues. So, for example, if openness and patience are virtues operative in restoration or assisted recovery, then a good restorationist will possess those character traits, and a good restoration will be one in which the design is not too rigid and has a time frame consistent with the rate of ecological processes. Third, among the appropriate goals of a good restoration is moral improvement—that is, reinforcing or developing (individual, social, and environmental) virtues. Good restorative practice encourages perspectives—or evaluative, affective (emotional), conative (desires), and practical dispositions—that are sensitive to environmental and social values and are conducive to their possessor living well (Higgs 2003; Light 2000a, 2000b). For example, it might encourage connection to a particular place or understanding of and appreciation for ecological complexity.

These relationships between virtue and good restoration have featured prominently in the ethics of restoration discourse. For example, Eric Katz's familiar criticisms of restoration are virtue oriented. On his view, restoration is problematic because it *expresses* and *encourages* the human tendency to dominate and control nature, as well as overconfidence in our ability to do so. It expresses and encourages a vice, *hubris*. Katz also appears to hold the view that those who engage in restoration tend to be *disingenuous* or else *insensitive* to natural value. They claim to be reestablishing something—naturalness and natural value—that it is impossible for them to put back, since their involvement entails that the product will be an artifact, which he considers contrary to naturalness (Katz 2000). This is a virtue-oriented evaluation of restoration.

However, as Eric Higgs and Andrew Light have well argued, it is not an accurate evaluation (Higgs 2003; Light 2000a). It may be, as Higgs puts it, that "intentionality courts hubris" (2003, 285) and its associates: domination, techno-fixes, anthropocentrism (in the sense that only human welfare is morally significant), and commodification. But restoration can, and often is, approached virtuously, with, for example, humility, ecological sensitivity, caring, cooperativeness, respect for nature, patience, a sense of restitutive justice, gratitude, and cultural, communal, and historical (including natural historical) sensitivity and engagement. Moreover, particularly when it includes communal involvement and participation, restoration can also provide opportunities to develop many of these (and other) virtues (Higgs 2003; Light 2000b; Throop, chapter 2, this volume).

For the remainder of this chapter I will refer to character traits that make for good restorationists and the cultivation of which are among the goals of restoration as *virtues of ecological restoration*.

Virtue, Restoration, and Global Warming

The central issue of this chapter can now be formulated more precisely: To the extent that our world is becoming relatively less stable and less predictable climactically and ecologically, what are the implications for the substantive content of character traits that promote value through restoration and that we ought to try to cultivate through restoration?

Openness and Accommodation

The additional uncertainties introduced by accelerated climate change reinforce the significance of dispositions opposed to highly control-oriented, domineering restoration. Our capacity to predicate and control the outcomes of our ecological interventions is likely to be reduced. Even greater *humility* regarding our ability is therefore justified. This is particularly so for interventions (including restorations) whose designs would take a long period of time to be realized (such as a farm land to woodland restoration), given the amount of ecological change possible over that period. In such cases, *patience* will also be crucial, since the recovery may not occur on the time frame we might prefer, and we may not know until well into the process what the product is going to be like. Here, virtue very much requires a proper temporal and spatial understanding of the "*scale* of the systems we think about and seek to manage" (Hirsh and Norton, chapter 16, this volume).

Similarly, *restraint* with respect to how strongly we impose our designs and desires on an ecological space will be crucial. The increased rate and unpredictability of ecological change associated with global warming means there is a broader range of possible (and plausible) ecological futures. Therefore it will be increasingly necessary to allow the recovering systems to take their own course in response to factors that we may not (and, perhaps, could not) have foreseen and ought not to try to control. Restraint is also necessary because under these conditions (of rapid climatic change and increased uncertainty) we are more likely to make mistakes and misplace resources, particularly if we try to execute detailed, less accommodating or flexible designs.

The need for humility and restraint extends beyond restoration to other types of responses to the challenge of adaptation. The difficulty with technological, control-oriented approaches to addressing climate change and its ecological effects, such as seeding the oceans with iron and releasing aerosols into the atmosphere, is that they depend for their success on our capacity to predict and control the effects of the intervention. We have not done this well in the past, and doing it well in the future is likely to be made more difficult by global warming. Technology must play a prominent role in our response to the challenge of adaptation. However, the ethical profiles of energy- and climate-related technologies vary dramatically. Some technologies address the cause of the problem (e.g., those that reduce GHG emissions, such as hybrid vehicles and wind energy technologies) and contribute to our adapting to the ecological realities we confront. Others, such as geoengineering,[2] involve further manipulation of ecological systems rather than changing our societal, cultural, and production systems. In an age marked by amplified ecological uncertainty, technologies that are more control oriented are likely to be less successful than those that are not, and technologies that are more interventionist into complex ecological systems are likely to be less successful and have greater unanticipated effects than those that are not. This is a straightforward function of complexity and uncertainty in dynamic and integrated systems. It is no different for restoration than it is for prevention and mitigation. The greater the complexity, unpredictability, possibilities, and uncertainties, the more flexible we must be in or ecological practices, processes, and goals if we want to be successful in promoting ecological integrity, and the more humility and patience are going to be required.

What these virtues (and others associated with them, such as *tolerance*) have in common is openness toward the ecological future of a place and accommodation of nature and natural processes in determining that future. Even (or, perhaps, especially) in restoration, where design and intervention are ineliminable, openness, pliability, and restraint are needed more than ever given the increasingly uncertain future associated with global warming. I have suggested only that the dispositions constitutive of them may be refined and their relevance to good restoration amplified in light of global warming.

Historical Fidelity

I turn now to another virtue standardly associated with restoration, *historical fidelity*. As I have used the term, "historicity" is an element of ecological restoration. An assisted recovery exhibits historicity (and is a restoration) to the extent that its product (or its goal) resembles or incorporates elements from a past ecology of the place. "Historical fidelity" (or historical faithfulness, perhaps) is not, as I use the term, an element or feature of an assisted recovery. It is a character trait. It is—roughly—to be disposed to value, desire, feel pleased about, and act toward accomplishing historicity in assisted recovery. It is to prioritize historicity in the goals of assisted recovery. Historical fidelity is thus a feature of some people, not of some ecosystems. Historical fidelity (though not always understood as a virtue[3]) is valued highly by some advocates of ecological restoration (e.g., Higgs, chapter 4, this volume). However, in the age of global warming, in which greater openness and accommodation to an increasingly uncertain ecological future are appropriate, the justification for historical fidelity as a virtue of ecological restoration is weakened.

One of the primary justifications for historical fidelity is that it helps to promote ecological integrity by setting goals and incorporating elements that are well suited to the site of the restoration, since they have obtained and thrived there in the past. Moreover, by grounding the goals of assisted recovery in (past) ecological realities of a place (rather than just our desires or visions for it), historical fidelity functions as a check on hubris. It is also associated with and encourages ecological sensitivity, humility, and restraint.

However, in the age of global warming, the ecological future is less likely to resemble the ecological past. According to the most recent report of the IPCC (2007)—which is now widely regarded as providing a conservative estimate of the likely magnitude of climate change and its impacts—even with probable climate trajectories (i.e., the reality of global warming), there are going to be (with at least "high" confidence) substantial increases in the rate of ecosystem transitions. Rain forest will very likely turn to savannah; and desertification, reduction in glaciers and ice sheets, losses of coral reefs, and flooding of coastal regions will also be very likely. There will probably be significant increases in species at risk of extinction (likely between 20 and 70 percent) and rapid changes in species ranges. The ecological impacts of global warming will be geographically differential—they will be greater in some places than in others. But, in general, historical ecosystems are going to be increasingly poor proxies for ecological integrity; and native (/invasive) is going to be an increasingly poor proxy for ecological beneficial (/ecologically detrimental). As a result, good assisted recoveries from the perspective of ecological integrity will in general resemble less strongly the systems or places prior to degradation, and too strong a commitment to historicity could become a form of ecological insensitivity to ongoing ecological changes.[4] Strong historical fidelity could increasingly involve imposing ourselves against ongoing ecological processes. It could amount to a denial of ecological reality. The implication for historical fidelity would seem to be that *prioritizing historical systems (and elements of those systems) in assisted recovery would have diminishing value (i.e., will be less conducive) to realizing the ecological (and moral development) goals of assisted recovery.* Therefore, the ecological integrity justification for historical fidelity being a virtue of restoration (or, more appropriately, a virtue of restorationists) is weakened. Or, perhaps more accurately, a weaker historical fidelity (i.e., a weaker commitment to historicity in assisted recovery) would be justified.

William Throop argues that even as the effects of anthropogenic climate change increase, those with "the virtues of humility, sensitivity, and self-restraint . . . will tend to adopt conservative restoration goals . . . that exhibit a high degree of historical fidelity" (chapter 2, this volume). For example, he suggests that, given climate change, restorationists should use genetic stock from the southern portion of a species' range, rather than look to a different species altogether. However, getting genetic stock from a place different than the site of the restoration is to incorporate less historicity. It is a diminished role for the history of the place. If this is the case systematically—that is, if historicity diminishes as a component of good restoration in general—then historical fidelity must diminish as a virtue of assisted recovery. Humility, sensitivity, and self-restraint would increasingly be in tension with historical fidelity. As Throop recognizes, the extent to which this will be the case is an empirical matter. It is for this reason that the claim defended here is conditional: to the extent that the ecological future less resembles the ecological past, the ecological justification for historical fidelity is diminished. However, as discussed earlier, given our current ecological and climactic trajectories, it does appear that the ecological futures of most places will be quite different from even their recent ecological pasts.

Another justification for historical fidelity is based on *natural value*— that is, the value of the continuity of ecological processes and the productions of those processes in virtue of their being free from human intervention and design. Let us assume that there is such value. Some (Katz 2000; Elliott 2000) have argued that (even if there is natural value) appeals to it cannot justify restoration (and by extension historical fidelity) because it is part of the concept of the value that it could not be recaptured by restoration. The need for restoration entails that the natural history (the basis of the value) has already been broken and execution of the restoration involves further human intervention and design. Therefore, even a restoration high on historicity cannot increase or create natural value.

In response, advocates for restoration and historical fidelity might argue that natural value is not an all-or-nothing matter (Hettinger, chapter 1, this volume). Natural value can be possessed by systems (and individuals) to greater or lesser degree based on the extent to which they are free (or their production was free) of human intervention (or the marks of human intervention and design). In this way, an assisted recovery high on historicity would have more natural value (because it bears less of a mark of human intervention and design) than one that does not; and historical fidelity might be justified (at least in part) because of its conduciveness to promoting natural value in assisted recovery.

However, even granting this (not-all-or-nothing) account of natural value, the relevant features of global warming reduce the extent to which it justifies historical fidelity. This is because even as it would be *in principle* possible to reestablish or increase natural value through restoration, it is *in practice* going to be more difficult to do. As discussed earlier, given the reality of global warming, assisted recoveries that aim for ecological integrity will involve less historicity. As a result, they will have less natural value. The implication for historical fidelity is that this justification for it—that it is conducive to promoting natural value—is diminished, since it will not promote it as much or as effectively. More intervention, not less, will be required to accomplish high levels of historicity.[5]

Alternatively, suppose that a previously degraded ecological place can develop natural value over time if it is free of human intervention, design, and control. On this conception of natural value, an ecological space can come to possess natural value proportional to the extent to which and duration for which it is free from human interference. However, even this conception of natural value does not support historical fidelity in a climate-changed world. The reason is that natural value emerges by allowing nature to develop independently. But to the extent that global climate change is ecological reality and the ecological future is accelerating away from the ecological past, realizing historicity will require greater intervention and control.

The distinctive features of global warming appear to weaken both the ecological integrity and the natural value justifications for historical fidelity (or favor a weaker historical fidelity). These are not the only justifications that have been offered for historical fidelity. However, they are the primary ones. Therefore, to the extent that the reality of global warming and the challenge of adaptation weaken them, the case in favor of historical fidelity as a virtue of restoration is substantially reduced.[6]

I want to try to be as clear as I can about what I am and am not suggesting. First, I am not claiming that ecological knowledge—for example, regarding the past ecology of an area or similar areas (or regarding global warming for that matter)—will be immaterial to designing a goal and developing a process that effectively realizes or promotes ecological integrity. I am making the weaker claim that, in general, appropriate ecological goals (designs) for assisted recoveries will bear less resemblance to past ecosystems of recovery sites. And, therefore, a weaker historical fidelity in setting goals and developing processes is justified.

Second, I am not claiming that in the age of global warming appropriate ecological goals—designs—for assisted recovery of a place will *not at all* resemble the prior ecology of that place. I am claiming that, in general, to the extent that the relevant features of global warming obtain, appropriate ecological goals will less strongly do so. Context-specific assessments of the role of historicity in assisted recovery are of course needed, and historicity will appropriately play a larger role in some recoveries than in others (Harris et al. 2006). Relevant considerations include global warming impacts on the area's climate (not all places will be equally impacted); the purpose of the restoration (e.g., research, memorial, habitat, or water purification); the desires (and values) of those involved; available resources; and the time gap between the degradation and the recovery. (For more on this, see Light, chapter 5, this volume.)

Third, I am not claiming that the historical continuity of ecological processes or the productions of those processes are any less (or differently) valuable given global warming. That is, I am not suggesting that the significance of natural value is somehow reduced by global warming.[7] I am claiming that, given global warming, reestablishing natural value through assisted recovery will (as a matter of ecological fact) be increasingly difficult to achieve, and therefore, appeals to natural value will do less to justify historical fidelity.

Fourth, I am not claiming that historical fidelity is completely unjustified or should be given up altogether. It is rather that the bases for historical fidelity—the commitment to having recovered systems resemble past systems—are weaker given global warming. To the extent that global warming is part of the ecological present and future, and to the extent that global warming involves increased ecological unpredictability and rates of change, historical fidelity, in general, is less conducive to accomplishing ecological integrity and reestablishing natural value, and is for that reason less justified.

Finally, I am not claiming that the justification for assisted recovery is diminished in the age of global warming. There is considerable evidence that assisted recovery of degraded ecosystems can increase substantially the biodiversity and ecosystem services of the system (Rey Benayas et al. 2009). What is at issue here is not the extent to which we ought to try to contribute to repairing ecological damage. It is rather the extent to which we ought to incorporate a commitment to historicity in our efforts to do so.

Given the definition of "restoration" that I am working with—that assisted recovery is restoration to the extent that it incorporates historicity—an implication of the diminished role of historicity and historical fidelity in good assisted recoveries is that assisted recoveries will, in general, be less restorations.

Reconciliation

The last character trait that I am going to consider is *reconciliation* (or, more properly, reconcilitoriness). Global warming elevates its importance as a virtue of ecological restoration.

Reconciliation, as I mean it here, is—roughly—a disposition to accept and respond appropriately to ecological changes that, though unwanted or undesirable, are not preventable or ought not be actively resisted. Reconciliation is not a commonly discussed environmental virtue (or virtue of restoration, for that matter). But it has, I think, always been relevant to ecological engagement and practice (and recognized as such). Even independent of global warming, ecosystems are always dynamic, and individuals, species, and abiotic features are always coming into and going out of existence. Good ecological engagement (for example, love and wonder toward nature) and practice (even preservationist activities) require accepting and not resisting (too strongly) such changes and losses.

Given the increased rates of changes and losses associated with global warming, the salience of reconciliation as a virtue of ecological practice (including in assisted recovery) would seem to be raised. Reconciliation with respect to global warming is not merely acknowledgment. It involves accepting the unwelcome fact of foregone global warming and its consequences. It is to have global warming inform our environmental perspectives, norms, and practices. It involves restraint toward ecological changes that cannot (or should not) be prevented, population or species losses that cannot be avoided, and aspects of the past that cannot be put back—even though they would not be occurring were it not for anthropogenic GHG emissions.[8] It also involves cultivating sensitivity and appropriate responsiveness to the value of biotic systems and living things that are the successors or beneficiaries of global warming.

Again, I want to be clear. Reconciliation as I mean it here is not resignation to global warming beyond what is foregone. That is, it is not contrary to working to, so far as possible, *reduce* further GHG emissions and *prevent* further anthropogenic climate change. Nor is reconciliation passive acceptance. It is not contrary to trying to *mitigate* the effects of global warming. It is to have those efforts appropriately informed by, or adjusted to, the ecological realities associated with global warming. Nor is reconciliation absolution. It is not contrary to culpability for global warming or remorse over what is lost. It is to accept the consequences and take responsibility for the activities that brought it about.

It involves recognizing that humanist adaptation—adapting ourselves to the world, rather than it to us (Thompson and Bendik-Keymer's introduction to this volume)—now requires accommodating the effects of foregone global warming. Pushing back futilely, inefficiently, or dangerously against those effects, trying to remake things as they were or otherwise would have been (out of a sense of guilt, responsibility, restitution, historicity, or nostalgia) is trying to remake the world—trying (yet again) to adapt it to us, rather than us to it.

Conclusion

I have tried to take seriously the idea that some anthropogenic global warming is part of the ecological present and future of the planet; and I have suggested several possible implications of this for *virtues of ecological restoration* (or perhaps, more aptly at this point, *virtues of assisted recovery*): the salience of virtues associated with openness and accommodation is raised; the justification for historical fidelity is diminished; and reconciliation will be crucial. These claims are premised on global warming increasing substantially the rate of ecological change and decreasing substantially our capacity to predict ecological change. They are justified only to the extent that this is the case. Thus, ultimately, whether these claims turn out to be true will depend not only on the reasoning behind them, but also (rather heavily) on the facts about global warming (i.e., whether and to what extent it increases ecological unpredictability and rates of change) and the facts on the ground (i.e., how these impact the effective practice of assisted recovery).

Acknowledgments

The author thanks Allen Thompson, Jeremy Bendik-Keymer, William Throop, Ned Hettinger, Philip Cafaro, John Basl, and Benjamin Miller for the helpful comments on earlier versions of this chapter.

Notes

1. See part IV of this volume for more on the point.

2. For more on how geoengineering intersects with moral questions, see Gardiner, chapter 12, this volume.

3. It is possible to translate the character trait of historical fidelity into a rule of historicity (and vice versa): prioritize historicity in assisted recovery. The discussion that follows regarding historical fidelity understood as a character trait applies, *mutatis mutandis*, to historical fidelity understood as a rule.

4. For example, in the Long Island Sound, lobster populations are declining while blue crab populations are increasing. This is due in large part to a few degrees' change in temperature and it is going to have significant ecosystem effects. Actively intervening to maintain the Sound as a fertile lobster habitat (or to try to maintaining the associated ecosystem features) is not feasible. Even on conservative global warming estimates, the Sound is going to increasingly suit blue crab rather than lobster, and this needs to be part of the ecosystem management (and expectations) for the place.

5. One possible response to this argument is that the importance of natural-historical value actually becomes greater (people value it more) given its scarcity and tenuousness given global warming. So even though historical fidelity is less conducive to promoting it, it is as justified due to the greater importance of the value.

6. Higgs (2003) argues that filling our nostalgic needs is a justification for historicity in restoration. If this is true, then in an age of global warming our nostalgic needs are increasingly in tension with the goal of ecological integrity. Our nostalgic needs—our desire for the past—could increasingly become nostalgic impositions on ecological integrity within the context of assisted recovery. As a result, our nostalgic needs are increasingly unsatisfiable (or more difficult to satisfy), or satisfiable in a less robust (or sustainable or difficult to sustain) form. Moreover, they are increasingly in tension with ecological sensitivity and humility, which are among the important moral education goals of assisted recovery. For these reasons, satisfying our nostalgic needs might well be another casualty of global climate change.

Higgs (2003) also suggests that historical fidelity is particularly conducive to developing connection to place, which is good in itself and conducive to promoting ecological integrity. However, there are other factors besides historicity that are conducive to developing connection to place. Many people's connection to a place seem to have more to do with *their* history with the place and the opportunities (recreation, relaxation, education, study, art) that it enables, than it does with the general history (or historicity) of a place. (For Higgs's analysis of historicity, see chapter 4, this volume.)

7. If the basis of natural value is that people value deeply places that are independent of human impacts and designs and in the face of global warming more people value this (or value it more deeply), given its increased scarcity or fleetingness perhaps, then the significance of natural value might even be greater in the age of global warming. Alternatively, if people give up on this sort of value (i.e., value it less), given the ecological realities, its significance would be diminished.

8. Take, for example, species extinctions. According to the IPCC (2007), even on the more modest global warming projections, it is foregone that many species (e.g., some alpine, arctic, and slow migratory species) will be lost due to global warming. We need to reconcile ourselves to that. Species come and go, and even when it is anthropogenic, sometimes one has to let them go (though, of course, not gladly, or without remorse or guilt) (Sandler 2010, forthcoming).

References

Aristotle. (ca. 350 BCE) 1985. *Nicomachean Ethics*, trans. T. Irwin. Indianapolis, IN: Hackett.

Elliott, Robert. 2000. Faking Nature. In *Environmental Restoration*, ed. William Throop, 71–82. Amherst, NY: Humanity.

Gardiner, Stephen M. 2010. Is "Arming the Future" with Geoengineering Really the Lesser Evil? Some Doubts about the Ethics of Intentionally Manipulating the Climate System. In *Climate Ethics*, ed. Simon Caney, Stephen M. Gardiner, Dale Jamieson and Henry Shue, 284–314. Oxford: Oxford University Press.

Harris, J. A., R. J. Hobbs, E. Higgs, and J. Aronson. 2006. Ecological Restoration and Global Climate Change. *Restoration Ecology* 14 (2): 170–176.

Hegerl, Gabriele C., and Susan Solomon. 2009. Climate Change: Risks of Climate Engineering. *Science* 21 (325): 955–956.

Higgs, Eric. 2003. *Nature by Design*. Cambridge, MA: MIT Press.

IPPC (Intergovernmental Panel on Climate Change). 2007. *Fourth Assessment Report.* http://www.ipcc.ch/publications_and_data/publications_ipcc_fourth_assessment_report_synthesis_report.htm (accessed November 30, 2011).

Katz, Eric. 2000. The Big Lie. In *Environmental Restoration*, ed. William Throop, 83–94. Amherst, NY: Humanity.

Light, Andrew. 2000a. Restoration, the Value of Participation, and the Risks of Professionalization. In *Restoring Nature: Perspectives from the Humanities and Social Sciences*, ed. P. H. Gobster and R. B. Hull, 163–181. Washington, DC: Island Press.

Light, Andrew. 2000b. Restoration or Domination? A Reply to Katz. In *Environmental Restoration*, ed. William Throop, 95–113. Amherst, NY: Humanity.

Rey Benayas, José M., Adrian C. Newton, Anita Diaz, and James M. Bullock. 2009. Enhancement of Biodiversity and Ecosystem Services by Ecological Restoration: A Meta-analysis. *Science* 325: 1121–1124.

Sandler, Ronald. 2007. *Character and Environment*. New York: Columbia University Press.

Sandler, Ronald. Forthcoming. *The Ethics of Species*. Cambridge, UK: Cambridge University Press.

Sandler, Ronald. 2010. The Value of Species and the Ethical Foundations of Assisted Colonization. *Conservation Biology* 24 (2): 424–431.

Vogel, Steven. 2006. The Silence of Nature. *Environmental Values* 15 (2): 145–171.

4

History, Novelty, and Virtue in Ecological Restoration

Eric Higgs

Restoration as potential and incipient, as seed not fruit, as data not deed, is ongoing and cumulative, inviting incremental enrichment by future generations.
—David Lowenthal, "Reflections on Humpty-Dumpty Ecology"

The science and practice of ecological restoration have thrived for several decades on the idea that historical knowledge anchors our judgments and practice. The approach has shifted from the idea of fixed reference points to more recent process-oriented configurations. All of this is poised to change. The intensification of anthropogenic environmental and ecological change[1] is moving the bar and the baseline: it is no longer clear what, if any, historical references are appropriate for restoration. Moreover, there are concerns that restoration as we know it may pass on, and certainly global processes of change challenge the local and regional focus of most restoration efforts. In this chapter I borrow from Throop's notion (chapter 2, this volume) of historical fidelity as a primary criterion that extends from his healing virtues (humility, self-restraint, sensitivity, respect for the other), and tow history farther from the brink of oblivion than Sandler (chapter 3, this volume). Inspired by Thompson's invocation (chapter 10, this volume) of novel virtues for a novel future (environmental responsibility), I suggest that historical fidelity (or, historicity, the quality of thinking historically) is a new virtue appropriate to a rapidly changing nature. A source of inspiration is my decade-long field project in the Canadian Rockies examining landscape change with repeat photography. I mingle these with my understanding of the emergence of hybrid and novel ecosystems.

Some brush clearing is required to understand what ecological restoration is, a subject that is doted on by philosophers and ecologists alike. The Society for Ecological Restoration International, the organization advancing the science and practice of restoration, stipulates this definition: "Ecological restoration is the process of assisting the recovery of ecosystems that have been degraded, damaged, or destroyed" (Society for Ecological Restoration International 2004). Restoration is a process (not a product) that involves human assistance (not control). *Ecological restoration* is a more general term than *restoration ecology* (science of restoration), and the adjective "ecological" signals both the focus (ecosystems) and process (integrative). *Nature restoration*, as preferred by Hettinger (chapter 1, this volume), suggests that restoration activity emphasizes only nonhuman practices, whereas the practice of ecological restoration suggests human participation and connection. Indeed, many restoration projects thrive with meaningful and modest human engagement (Higgs 2003).

Over the last decade, three large changes have been under way in how we understand restoration. First, it is now almost universally accepted that successful restoration depends on incorporation of social, cultural, political, economic, and aesthetic practices (Parks Canada 2008; Higgs 1997). It is not necessary for human interests to trump ecosystem functions, and under the right conditions people are enlivened by restoration and are more likely to act as guardians of ecological integrity. Second, there is greater awareness of cultural variation in how restoration is understood and practiced. This is especially true in the global South, where significant traditions of subsistence and small-scale agricultural production are incongruent with wilderness-based North American notions of restoration. As Hall demonstrated in his comprehensive comparative study of restoration in Italy and the United States (2005), Europe's predominantly cultural landscapes admit a different kind of restoration than found in North America. Third, there is growing recognition that rapid environmental and ecological changes will force the meaning and practice of ecological restoration to shift. Just how far we need to go in modifying or displacing ecological restoration as a critical environmental practice hinges on the significance of history and also the extent to which we are willing to entertain flexibility in ecological futures: contingency in restoration admits a detachment from history.

Albert Borgmann's *Crossing the Postmodern Divide* (1992) proposes a distinctive quality of contemporary life in which we oscillate between sullenness and hyperactivity in responding to large-scale social and environmental challenges. Perhaps nowhere is this more evident than in the discussions on climate change. For two decades scientists and activists have pressed for decisive responses from individuals, governments, and corporations. The response has been mixed, but primarily sullen. This shifted in response to a cascade of stories of climate-related effects, burgeoning scientific literature demonstrating myriad climate processes and impacts, and the public sanction of the issue (especially in the wake of the 2007 awarding of the Nobel Peace Prize to the Intergovernmental Panel on Climate Change [IPCC]). Now the mood is decidedly hyperactive.[2] The United States under the influence of new leadership has unleashed a barrage of funding for innovation, science, and policy, although resolute national legislation remains elusive. Globally, there is increasing resolve to craft a serious international policy mechanism (the late 2010 negotiations in Cancun, Mexico) in spite of shortcomings at the 2009 United Nations conference on climate change in Copenhagen. In my home province of British Columbia, the government created in 2008 the largest endowment in Canadian history to establish the Pacific Institute for Climate Solutions. Change is in the air, literally and metaphorically. This hyperactivity is encouraging, and on the surface is more appealing than sullenness. The problem that Borgmann identifies is one of oscillation: how does one attenuate concern and resolve? This matter is important for ecological restoration. A fevered commitment to wrestling with rapid environmental change may displace sober assessments, and more specifically compromise core principles in restoration such as historicity.

Such oscillation frustrates the development of a robust commitment to living sustainably. Gathering traditional virtues in the service of change is promising, but what if, as many authors in this volume ask, we need a new way of thinking, new virtues? Hirsch and Norton (this volume) propose a change in our context to grapple with appropriate intellectual virtues, or indeed to evolve a set of virtues that respond to emerging conditions. In the manner of Aldo Leopold, who famously shifted from a resource management worldview that focused too much on individual species, to "thinking like a mountain," we, too, need to adjust our context to one that recognizes new models of responsibility and adapt institutions to serve the promise of "radical hope" (Thompson, this volume). Leopold was also a leading force in ecological restoration. He spurred early prairie restoration projects at the University of Wisconsin-Madison Arboretum, several of which are flourishing after almost eighty years, and conducted personal projects that restored land north of Madison (Meine 1991). This was toward the end of his life, years after the "dying green fire" incident that was pivotal in his shifting values. Thus, restoration was a practice accelerated by progressive mid-twentieth-century thinking that is now buffeted by a new climate, globalized mixing of species, and a digital temperament. This points perhaps to a capabilities approach suggested by Keulartz and Swart (chapter 6, this volume) that refigures our moral commitments to species and ecosystems in transition. Will novel ecosystems need "novel forms of human goodness?" (Thompson, chapter 10, this volume), including a new virtue that embeds our growing commitment to global responsibility, as well as the tempering virtue of history?

Landscapes of Change

In 1996 I followed a small research detour that opened a major new trail in my research. Passionate about mountain landscapes, I had begun work in Jasper National Park, one of Canada's iconic wild places, with a group of social and natural scientists. The ecological integrity of this 11,000-square-kilometer protected area was threatened by resource development on the outside and visitation and development on the inside. My particular contribution to the growing synthesis was that any determination of future landscapes depended on knowing something about past ecosystems and the changing values that had shaped them. Forest ecologists computed history through tree rings. Paleoecologists interpreted deeper time from lake and wetland sediments. Historians sifted the archives, and anthropologists examined prehistory and oral histories. In assembling this knowledge, I was made aware of a remarkable collection of systematic photographs taken from the tops of dozens of mountains in Jasper. Eventually, Jeanine Rhemtulla, then a graduate student, and I climbed all ninety-two peaks in what we learned was the 1915 phototopographic survey of north-central Jasper. The 735 photographic pairs are remarkable portraits of change. Patchy forests burned over in the 1880s succeeded to close-canopy ecosystems. River courses changed. Glaciers retreated. And, throughout the park scant evidence of human activity in the historical imagery gave way to highways, railways, trails, outlying accommodations, encampments, a town of five thousand, and utility corridors (Rhemtulla et al. 2002; Higgs 2007; MacLaren, Higgs, and Zezulka-Mailloux 2005).

My fascination with historical ecology was stoked by those initial climbs. The initial photographic collection in Jasper was a drop in a large bucket. From the 1880s through the late 1950s, national and provincial surveyors mapped most of Canada's mountainous regions (primarily in Alberta, British Columbia, and the Yukon, including the Alaska-Yukon

boundary). This collection is the largest of its kind in the world, constituting more than 140,000 photographic negatives on glass plates, photographic prints, maps, notes, and journals. The sheer effort of coming
to understand an archive of this scale has become a major new scholarly
trail, and has become the Mountain Legacy Project (www.mountainle-
gacy.ca). Through it I grew preoccupied with dynamism in these ecosystems, that everything was in flux (figure 4.1). The more I peered through
the lens of the various cameras we took with us, the more uncertain I
became about what constituted an appropriate reference point. After all,
the forested ecosystems of the late nineteenth century after several
sequential massive fires were as exceptional as those following fire suppression during the twentieth century. Having taken more than a thousand repeat images, and supervising the repeating of several thousand
more, there are some lessons.

    First, the dynamism of ecosystems at all scales is difficult to overstate,
and at the same time difficult to comprehend. Visible change rendered
at a scale sensible to human perception comes often in the form of
episodes or catastrophes such as wildfire or flooding or blowdowns.
Second, gathering historical ecological information for reference purposes is a painstaking and difficult matter involving extensive and sometimes expensive fieldwork. As historians can attest, seldom are stories
of the past packaged neatly for analysis and interpretation. Arriving at
explanations that meet robust conditions of explanation and can be
challenged by others cannot be accomplished easily. For example, graduate student Will Roush discovered using historical and repeat imagery
that the upper reaches of subalpine larch (*Larix lyallii*) in the Goodsir
basin of Kootenay National Park has shifted upward, and not just
upward, but also more dramatically than observed for almost all other
trees species globally. Did this ecological change emanate from compo-
nent-driven processes at the micro- and mesosite scales, or was it shaped
by constraints imposed by climate? Two summer and two winter seasons
of painstaking fieldwork involving comprehensive ecological measure-
ments yielded a reasonably clear answer: climate is the stronger signal
and likely significantly responsible (Roush 2010). And the third lesson:
the ardor of such projects brings unexpected benefits. The very process
of gathering historical references introduces the observer to myriad patterns, processes, and structures in the ecosystem that are revealed through
systematic study. Ian MacLaren studied the people who were displaced
by the creation of Jasper National Park in 1907. Of mixed indigenous
ancestry, the Métis peoples homesteaded and farmed the Athabasca

Figure 4.1
View of the town of Jasper facing north from The Whistlers Mountain. The top photo from 1915 is by M. P. Bridgland, and the repeat image (below) is by J. Rhemtulla and E. Higgs from exactly the same location in 1998. The build-out of the town is striking, as are the changes in vegetation.

Valley in the latter half of the nineteenth century. Until recently little was known of their farms, and even less about their role in the landscape (how can farming be consistent with a wilderness landscape?). By zooming in on one of the historical photographs rendered in digital form, MacLaren noticed the farmsteads, fences, and fields (MacLaren 1999). The same examination of the repeat image showed significant vegetation encroachment, making the historically open landscape that much more striking and significant.

The type of history articulated by the Mountain Legacy Project does not render deep time. Context-dependent, deep time refers in my case to documentary evidence of past climate and ecological/cultural conditions before any known records (in Jasper, before 1800). Dendrochronology takes us back a few hundred more years, encompassing the Little Ice Age (approx. sixteenth–nineteenth centuries), but typically does not provide a detailed glimpse at the preceding Middle Warm Period (Luckman and Kavanagh 1998).[3] Excavation of lake and wetland sediments and their associated pollen and charcoal records permit a glimpse into the farthest reaches of the Holocene and in some cases just at the recession of the last major glaciation (Hallett and Hills 2006). The record provides clues of past conditions in which forests were once grasslands, sea levels were lower (or higher), and some ecological communities fared better in a location than they would today. Notwithstanding rapid anthropogenic climate change, the pace of temperature and precipitation changes, and their interannual variation, was sufficient to allow ecological processes to keep pace with environmental change. From a restoration point of view, one could plan for past conditions, especially if average temperatures and precipitation values were roughly similar to today. An adaptive approach would allow the ecosystem to follow one of several alternative trajectories with further interventions to control weedy species or benefit target (e.g., threatened) species. The work of historical research, therefore, was of obvious benefit in finding a range of appropriate reference conditions (White and Walker 1997), and thereby provides a clear set of goals for restoration.

Hybrid and Novel Ecosystems

Now we have the certainty, or rather growing certainty, that environmental changes will exceed our practical ecological reconstructions of the past, which generously I gauge as the Upper Pleistocene and Holocene epochs. This is the interval when people radiated around the globe and created the records that have become part of contemporary cultural practice and awareness. For instance, indigenous people in Australia trace their origins roughly forty thousand years BCE, and have a continuous record through various kinds of cultural transmission. Suggestions for moving beyond the Holocene in North America in setting goals for restoration have drawn considerable response. For example, there are proposals for inserting a missing large vertebrate species (e.g., wooly mammoth) with a conspecific (e.g., elephant) to fulfill missing ecological functions; this pushes the boundaries of conventional restoration thinking (Donlan et al. 2006). Nonetheless, the notion of using reference conditions as homologous in the present era begins to weaken in the face of conditions outside the known range of historic variation. However, what if we adopt a notion of historical references as *analogous* instead of homologous? Would this weaken the historicity of ecological restoration, or simply change the role of history? Does this mean that *historical fidelity*, a term I helped popularize in the mid-1990s (Higgs 1997), fades in significance?

Evidence is pointing to significant adjustments in ecosystem composition and structure in response to changes in environmental variables. These are not simple changes and they will not be experienced as strictly linear. Temperatures will rise overall, but polar regions, for example, will show more significant changes. There will be episodes of intense warming that will break records, and years when weather seems relatively benign. Patterns of precipitation will change, and for some regions these will prove difficult to anticipate. Severe events will rise in both frequency and intensity (IPCC 2007). Cascading shifts in other environmental variables will cause unexpected changes in ecosystems. Species invasions will rise, and produce uncertainties in how many resources to dedicate to manage such changes. For example, alterations in nutrient cycling will cause systems to move closer to or pass over thresholds. The heterozygosity of some plants will allow for significant resilience; others will struggle with modest changes. Confounding our understanding of changes is the heterogeneity of ecosystem responses to environmental change at different scales. Subtle landform differences will produce distinct local responses. For instance, the southerly aspect of a slope will typically show a more pronounced and rapid response than northerly slopes. In urban areas, people will use technological means (irrigation, shelter) to assist particular species and assemblages. The type and intensity of people's adaptive responses will matter, too. The fact that change will happen does not mean that any change will be accepted: strident interventions will preserve or slow the changes in particular ecosystems that provide significant ecosystem services, including cultural services.[4] Finally, while the changes are linear, we will experience them episodically because of hysteresis and various regime shifts. Thus, uncertainty about the qualities of change and adaptive responses to them is very great, and inversely related to the certainty of a shifting climate.

A growing realization of the combination of environmental (climate) and ecological (species invasions) change is producing ecosystems that are manifesting properties outside of historical referents. These go variously by terms such as non-analog or novel ecosystems, and point to systems that may or may not bear resemblance to ecosystems we have experienced in modern times or indeed that we are aware of from historical and prehistorical records (Harris et al. 2006). In southwestern Australia, former eucalyptus woodlands predominated. Agricultural development and decline have produced novel ecosystems in old fields. Thus, a combination of land use legacies including secondary salination, reduced native seed sources, altered soil chemistry, invasive species, and shifting climate are producing alternative and relatively robust ecosystems resistant to traditional restoration techniques. The effort required to restore such a system is heroic, and the opportunity for reinvasion very high (Jackson and Hobbs 2009; Hobbs, Higgs, and Harris 2009; Hobbs and Cramer 2008).

Novel ecosystems receive most attention, but there is an important intermediate category between novel and historical: hybrid ecosystems. Hybrid ecosystems are admixtures of historical features and new players. Locally, in southeastern Vancouver Island, invasive species such as Scotch broom (*Cystisus scoparius*) and English ivy (*Hedera helix*) have transformed scrub, meadow, and forested ecosystems. However significant these invasions, the native tree flora remain significant. Ground flora are suppressed, but the seed bank awaits restorative action. Systematic restoration efforts can shift the systems toward historical referents, but the tide is always pulling the ecosystem back toward hybridity. The likelihood of restoration success is tied to the extent of environmental, ecological, and social change (Hobbs, Higgs, and Harris 2009). Decisions are made regularly regarding how much effort is appropriate for specific ecosystems, and the calculus is based on site conditions, ecological and cultural significance, political and economic support, and mobilization of volunteers. As the tide pulls toward novelty, the decision to restore will be increasingly difficult.

The Functions of History

Historical fidelity and more generally historical qualities have always been of concern in ecological restoration, and have come under theoretical investigation in the last fifteen years (Hall 2010; Egan and Howell 2001; White and Walker 1997). In earlier work, I characterized historical fidelity as the quality of respecting the historicity (the quality of being historical; historical authenticity) of an ecosystem (Higgs 1997, 2003). The very notion of restoration embodies historicity, although in ecological terms there are significant departures from practices such as architectural restoration, in which compositional fidelity matters more than processual fidelity. Suppose one purchases a rundown mid-twentieth-century architect-designed home that despite years of neglect displays a remarkable embodiment of modernism in house design. The first challenge is in deciding how much to honor the original design, or whether the better choice is to remake the house to contemporary interests. Supposing the former, the next choices involve the degree of faithfulness by which the restoration will proceed. Will the original layout be conserved or recovered? Will period fixtures be sought? Will windows and other structural features be updated? How far is it appropriate to go in restoring the house? Of course, the answer depends on many practical considerations, including the significance of the original design (how rare an example is it?), the resources available for restoration, and the intent of the new owners. No matter the choice, the goal is easily resolved by focusing on the original qualities of the home. Thus, the composition of the house matters most in ascertaining the goals of the restoration. Typically, this is not the case for ecosystems.

The striking difference between ecosystems and created objects is that ecosystems are internally dynamic entities.[5] Ecosystems are typically open to exchanges of energy and matter and are self-regulating. Myriad processes such as plant succession, pollination, water currents (in aquatic ecosystems), changes in seasonal temperatures, and severe wind events alter the ecosystem continually; there is nothing static and therefore nothing fixed about an ecosystem that would allow clear or simple historical goals. Thus, the determination of historical references under conditions of relatively stable or slow-changing environmental conditions is complicated by an ecosystem's dynamism. Reference conditions are used as a guide to setting goals (Egan and Howell 2001; White and Walker 1997), which are seldom concerned with returning a degraded ecosystem back to the way it was in the past. More realistic is an acknowledgment that removing the sources of degradation and restoring historical pro-cesses, patterns, or structures (or some combination of these three) open up new adaptive and evolutionary pathways (Parks Canada 2008). *Resetting* the ecosystem may be an apt metaphor for much of what is conducted as restoration presently, although in practical terms an eco-system cannot be reset in the way a relatively closed electronic system can be. The idea is that historical qualities bear on setting goals for res-toration, and new environmental conditions, including those shaped by people's activities, influence the further development of the ecosystem along one of several potential trajectories (Suding and Gross 2006).

What has complicated the discussion of historicity and ecological restoration is the variety of types of historical explanation. There is a tendency to think of one type of history (e.g., restoration returning a system to a past condition), and I suspect that notions such as historical fidelity have abetted this view. In fact, there are several varieties, each of which presents restoration in a slightly different light and all of which can operate simultaneously:

1. The first and perhaps most prevalent is history as a means for redress-ing specific disturbances. The goal of restoration is to return the ecosys-tem to its proper functioning condition (often including structural and pattern-based characteristics) prior to a clearly identified disturbance. Suppose a dam altered fish migration, as so many have. Restoration would remove the dam to allow a prior condition. The value of historical knowledge is in setting objectives.

2. Restoration of ecosystem as artifact mimics the process of restoring buildings or works of art. It is relatively rare, such as the restoration of the battlefield at Antietam to its American Civil War condition in 1862. Included in this restoration is the "replanting of historic woodlots and orchards" (United States National Park Service 2010) to provide a real-istic version of the landscape at the time of the battle. While relatively rare in practice, this notion of ecological restoration led to a widespread misunderstanding of restoration as returning ecosystems to exact condi-tions on a particular date (e.g., 1491, just prior to Columbus's landing in the New World).

3. Recognizing the dynamism of ecosystems, a historical range of varia-tion (HROV) is presumed over a delimited interval, and the goal of restoration is to move the ecosystem back within the bounds of various discrete parameters (Egan and Howell 2001). In Waterton Lakes National Park, where I worked, a challenge was in returning disturbance processes (fire, mainly) to the Front Range prairie/aspen matrix. In the absence of

fire in the twentieth century, aspen trees had encroached rare native grasslands. This approach has proved popular as a means for acknowledging a moving ecological target, and to decenter restoration practice from aiming for particular episodes (Levesque 2005).

4. Closely related to the HROV approach is the idea of resetting an ecosystem to a condition of greater health or integrity. Typically, and by definition, an ecosystem is more integrated at some time in the past. The subtle difference with resetting an ecosystem is that the focus is less on historicity per se than with ascertaining ecological performance (e.g., species richness, invasion resistance). This approach also acknowledges that ecosystems may veer off in one of several directions, all of which are appropriate given shifts in environmental and ecological conditions. The hand of the restorationist is restricted to resetting historical qualities. The question of ongoing intervention (e.g., selective thinning or weeding of invasive species) is left open.

5. History also provides references for understanding how ecosystems have functioned in the past, and how they might operate under new conditions (e.g., elevated temperature, more human activity). Reference in this sense does not bind the restorationist to a particular course of action. Much like the use of references in writing a book chapter, ecological references inform our desired interventions. For example, it may be valuable to know the historical hydrology of a landscape that is about to be developed for housing, but this does not bind the developer to restoring the stream condition to the way it was; the option may exist to return a stream feature to the property that gestures to historical referents. Another common example is the disturbance profile (e.g., mean fire-return interval) of fire-dependent forested ecosystems that provides vital information in setting restoration goals.

6. We use common understanding of historical human practices and beliefs in determining restoration goals. This is history as postcolonial reconciliation, wherein our motivations for restoration are tied to redress of damages wrought by colonial disruptions of traditional lifeways and ecosystems (this includes the damage done by highly invasive exotic species). In the city of Victoria British Columbia, where I live, the fervent restoration of relict Garry Oak savannahs is partly motivated by that the sense of responsibility for damage created (Vellend et al. 2008; MacDougall, Beckwith, Maslovat 2003). The goal of these restoration projects is recovery and conservation of native species, but also the regeneration of traditional ecological management practices (e.g., harvesting of traditional native root crops, frequent light fires, weeding).

7. Historical knowledge also serves as a governor on our exuberant ambitions. In the same sense that a governor works on mechanical systems to limit the maximum output or speed, history reminds us to exercise caution and careful choice in how we restore. The very act of researching the historical conditions of an ecosystem slows us down to consider the longer-term changes that an ecosystem experiences, and how our contemporary intervention will likely play out. This particular function of history will become especially important under conditions of rapid environmental change.

Together these seven types of historicity complicate the meaning of historical fidelity, expressing variations in the exactness and value of historical referents. These characterize how we use history, and each one represents particular values that give privilege to historicity and historical fidelity. In *Nature By Design* (Higgs 2003), I proposed that three features of ecological history made it central to the work of restoration, and the absence of historical insight rendered restoration as something else entirely (reclamation? revegetation?). First, we long for the past because it gestures to a time when ecosystems had greater autonomy and integrity. We do not long for it in a simple way, and our view of the past is partly obscured by a thickly coated cultural lens. I called this *nostalgia*, in the sense of a "bittersweet longing for the past," to reflect that we want something that is beyond our easy grasp. The problem is that nostalgia amplifies, sometimes distorts, the contingency of historical knowledge (i.e., partial, relative meaning). To nostalgia I added the principles of *narrative continuity* and *rarity*, both of which contribute to adding value to historicity. Narrative continuity describes the process by which we seek identity with a place through the retelling of stories. We create relations with a place by linking ourselves through time to the unfolding of events. Sometimes the stories tell of change (e.g., retelling the story of a hike to Angel Glacier on Mt. Edith Cavell in Jasper National Park that is now no longer recognizable as an "angel"), and others emphasize permanence (e.g., returning to a giant Douglas fir in the Carmanah Valley that seems unchanged since a first visit years before). These retellings, irrespective of the manner in which they are delivered, bind us to the historical qualities of an ecosystem. The third quality is rarity, which increases the value of an ecosystem. Those that are rare in an ecological sense (e.g., coastal sand dunes on southern Vancouver Island) or a remnant coastal Douglas fir ecosystem in the city of Victoria, carry a special value that warrants greater attentiveness to conservation, preservation, and restoration.

In identifying new varieties of history, new sources of value are added. Nostalgia, narrative continuity, and rarity cover the first five varieties reasonably well. We are motivated, for example, to focus on historical range of variation (3) because this tells an effective story about how ecosystems change over time, and harkens back to memories of ecosystems that express our longing for them to be restored. Rarity is a powerful motivation in restoring artifactual ecosystems (2), ecosystems so valuable that we wish to freeze them in time. The last two varieties of history (history as postcolonial reconciliation; history as governor) generate new sources of value for restoration. To seek redress for the damage wrought by colonizing peoples and other species (rapidly invasive and exotic species) we tap into the value of *reconciliation.* To set limits on our capacity to act incautiously, we invoke the value of *restraint.* How might we translate these sources of value into virtues appropriate for intervening responsibly in a rapidly changing world?

Novel Virtues?

Increasing rates of environmental and ecological change alter defining variables for ecosystems: temperature, precipitation, storm events, and so on. This creates nasty problems for our embrace of historicity in ecological restoration. Certainly several types of history described earlier recede in use. We cannot, for example, redress specific disturbances by going back even a few decades to a prior state. Or, can we restore ecosystems as artifacts without heroic interventions (such heroism might be evident as highly valuable ecosystems are restored and preserved in the same vein as zoological and botanical gardens). Resetting ecosystems is also tricky, especially when the environmental conditions have changed beyond the point where conventional historical referents are feasible. A reasonably tidy and fashionable notion was that if ecosystems operated within some reasonable long-term historical range of variation, for instance the Holocene epoch, then all is well. However, if the range exceeds our memories and our proxy data, then the references begin to pale. Hence, Sandler (this volume) proposes we find another term to describe our work in "assisted recovery," such as *reconciliation* (see also Tucker 2005). Thus, our conventional historical touchstones for restoration and the values that elevate history are swept away.

Or, are they? Nostalgia and rarity are likely to intensify as ecosystems shift outside our range of conventional appreciation. Certainly the values of reconciliation and restraint will be amplified. The problem will come in narrative continuity. Hybrid and novel ecosystems (Hobbs, Higgs, and Harris 2009) will make ecosystems unfamiliar and in some instances discontinuous. Ecological interventions will produce ecosystems that will be more like simulacra (copies with no true original) than faithful restorations (Cypher and Higgs 1997). Thus, if we take a multivalent approach to understanding history in ecological restoration, and ecological intervention more generally (Hobbs et al. 2011), we find there are significant sources of historically significant values that remain.

This is consistent with Throop's argument (chapter 2, this volume) that historical fidelity is a conservative approach to restoration consistent with the four key virtues of his healing metaphor: humility, sensitivity, self-restraint, and respect for the other.[6] Humility intercedes against hubris, sensitivity compels us to pay close attention to ecosystems and not just our own desires, self-restraint places a brake on our ambitions, and respect for the other attunes us to that which is other than human. Throop's argument is especially important in responding to the unpredictability of future ecological and environmental changes. What seems likely is that we will confront problems that we have not previously known, and that our responses are best when they are cautious and circumspect. Knowledge of history exposes clues about what might happen, and the mere study of history forces a better understanding of a particular ecosystem and ecological principles and functions more generally. Throop is also concerned that removing the condition of historical fidelity opens up "a version of gluttony" (consistent with history as governor). A hyperactive large-scale response that focuses on ecosystem services at the expense of historical fidelity and ecological integrity could look a lot like gluttony.[7]

Ron Sandler asks: "what are the implications for the substantive content of character traits that promote value through restoration and that we ought to try to cultivate through restoration?" (chapter 3, this volume) A decreasing capacity to predict outcomes of ecological restoration should produce greater humility. Patience will be required to await the outcomes of our actions. Restraint is needed to counter a desire to continue interventions instead of allowing ecosystems to find their own course. Taken together, such virtues acknowledge openness to potential futures and accommodations needed to account for less certain outcomes. For Sandler, a rapid changing environment does not change the virtues associated with restoration, rather the virtues are adjusted to reflect greater uncertainty and volatility.

The virtue of historical fidelity, however, falls prey in Sandler's version, to a warming climate.[8] He proposes a distinction between strong and weak historical fidelity. The strong version, associated with traditional restoration, takes robust and accurate cues from past ecosystem states for defining restoration goals. The weaker version, made necessary by climate uncertainties, suggests that historicity is valuable to the extent that it supports the recovery of ecosystems, and does not interfere with the flourishing of an ecosystem under significantly new conditions. The erosion of historical fidelity is tied partly to the diminishment of natural value. New conditions and novel ecosystems will lessen the natural value associated with historical elements.[9] Sandler is clear that his arguments do not deny historicity; instead appropriate ecological goals in the future will rely less on historical referents. What is needed is a new stance to ecosystems, one that depends crucially on reconciliation and not restoration.[10]

Taken together the arguments in favor of maintaining a grip on historical fidelity are considerable. The prudential and practical consideration of historical referents urges us to seek lessons in the historical record. These lessons are useful both to the extent they provide clues about restoring ecosystems and also entreat us to dig deeper into an understanding of how ecosystems function. The warrants of nostalgia, narrative continuity, rarity, reconciliation, and restraint still hold value in a rapidly changing world, although some (i.e., narrative continuity) do inversely in relation to the magnitude of change. The question that Sandler exposes by invoking the idea of reconciliation is: Will rapid environmental and ecological change force us to develop a new approach to ecological restoration, or will a renovated version continue to hold true as a general practice? In other words, do we abandon history, or simply rely on it less intensively?

Historicity is one of two moral centers of gravity for ecological restoration (the other is ecological integrity). What restorationists ought to do is conditioned by what was. Variously formulated, historical knowledge plays a crucial role in shaping our response to ecosystem damage. This paradigm grew ferociously in the last half of the twentieth century, becoming one of the dominant factors in ecosystem-based management. A few years into the twenty-first century, its dominance is challenged by rapid environmental and ecological change. Legitimate concerns over the development of novel ecosystems compel new approaches. Restoration plays a role, an important one, but perhaps not a dominant one in the future of addressing human degradation of ecosystems.

Historicity is consistent with and promotes traditional virtues that Throop, Sandler, and others argue are important for ecological restoration, and more broadly as a way of enlivening beneficial human agency in relation to natural processes. In this respect it constitutes what Thompson (this volume) describes as a "novel virtue," one that is critical in navigating a changed world. He argues that "environmental responsibility" is a new virtue required for a sustainable way of living: it "is a disposition to address well the demands on one's character that the role of stewardship entails; it is a character excellence regarding the possession of other specific virtues of stewardship, such as benevolence and loyalty." Historicity, then, becomes a candidate for a virtue in a world that is moving more quickly than we can easily know.

If historicity is central to the practice of these virtues, or in fact a novel virtue of the future, then ecological restoration in some altered form is what should be fortified and promoted. Work is under way recently in understanding the quality and implications for novel ecosystems, and the extent to which a broader form of ecological intervention is required as distinct from ecological restoration. It is difficult to gauge how far this departs from conventional norms, norms that will emphasize ecosystem services in distinction to historical ecological qualities and trajectories, or in other words, a more fully engineered approach. Should we end up displacing conventional ecological restoration, with its dependence on ecological integrity and historical fidelity, then historicity as a virtue, responding as it does to many varieties of historical understanding, will remain important. There's always the possibility, too, that the future will be so different that the past conditions of ecosystems will have the same value as historical relics do: rare glimpses of something from a dimly knowable era.

The debates over what virtues are appropriate to our adaptive responses matter, and precisely because they explain why a presently unfashionable virtue such as historicity is important. The bonds of affection that link us to the places we inhabit or remember are also the ones that contribute to virtues of humility, sensitivity, self-restraint, and respect for the other. It is less clear that such virtues will be preeminent in a world beyond restoration, where the practices are anchored by our desire to limit change through whatever ingenious and hyperactive responses we can summon. I learned through studies of landscape change in mountainous ecosystems that prudent action considers the ecosystems carefully, which includes fine-grained historical understanding. There are many models of history—I provided an inventory of seven—that offer guidance of different kinds for restoration or adaptive interventions. Even in cases where history is largely erased, the discipline of understanding the erasure enlarges our humility, sensitivity, and self-restraint. Historicity serves in applying brakes to unfettered interventions in ecosystems.

The practice and conceptual basis for ecological restoration will change, and should change, but all the while the value of historicity will play an important role. How far will this go in a changing world? This depends, of course, on the rate of change and our collective capacity to maintain civil society. I have argued (Higgs 2003) that "restoration" is not the best term to describe our beneficial engagement with ecosystems, and that terms such as "regeneration" reflect the reality of change more effectively ("reconciliation" is also an attractive candidate). "Restoration" is, however, the term that describes our present practice, and may have the medium-term tactical advantage of forcing us to stare more closely at history. Let's wait a decade or two before we switch allegiance and begin tearing out the foundation stones upon which the progressive practices of ecological restoration are based.

Notes

1. Environmental change refers to the broad suite of consequences of biogeoclimatic changes resulting from anthropogenic climate change. Ecological change is associated with economic and social processes of globalization, and is manifest primarily in increased rates and significance of biological invasions.

2. However, as I write this (early 2010) the mood of the public is shifting in response to a variety of controversies about the IPCC. We may well be heading for another interval of sullenness as climate change skepticism throws cold water on regulatory and governance changes.

3. Such broad labels describe intervals that are spatially and temporally variable, and manifest in slightly warmer or cooler temperatures. Precipitation differences were also significant in affecting tree establishment and glacial advance/retreat. And there are multiple dimensions to temperature: winter minimum, summer maximums, and so on, which play a crucial role in ecological formation and change.

4. A shift to thinking of ecosystems in terms of services will likely promote optimization of particular services. For instance, an urban wetland that offers important water recharge and detention services can be engineered for greater effectiveness, which will pull away from historical function and composition. However, aspects of ecosystems resonate powerfully with people and these cultural services will dampen tendencies for ecosystem manipulation and optimization (e.g., the Garry Oak ecosystems in Victoria, British Columbia).

5. So, too, are some created objects, such as buildings designed to change to suit emerging needs.

6. Throop and I disagree over the appropriate dominant metaphor for restoration. His use of healing is a congenial one for many, and follows a path worn by some restoration practitioners and scientists (e.g., William Jordan). In some respects, the debate ends up in the same place: both of us place value on similar virtues (humility, sensitivity, and self-restraint). My argument for design hinges on the recognition of our roles as interveners in ecosystem processes. The challenge is how best to intervene. Studying the evolution of design practice is to realize that design has shifted from an industrial practice to one of integrated work with complex systems. Designing ecosystems is the self-conscious act of recognizing and respecting their qualities, and working toward ends that meet ecological goals (as processed through human understanding).

7. Compare with Kawall, chapter 11, this volume, on greed.

8. Sandler converts historical fidelity to a virtue associated with restoration practitioners who are disposed to value historicity in restoration.

9. While I appreciate the consistency of Sandler's argument on natural value, the implications for restoration are potentially misleading. Constraining natural value to exclude long-term sustainable cultural practices not only eliminates many ecosystems from consideration, it also provides a skewed view of what is valued in restoration.

10. Presumably this approach to reconciliation draws its emphasis from the work of Rosenzweig who advocates a "science of inventing, establishing, and maintaining new habitats to conserve species diversity in places where people live, work, or play" (2003).

References

Borgmann, Albert. 1992. *Crossing the Postmodern Divide.* Chicago: University of Chicago Press.

Cypher, J., and E. Higgs. 1997. Colonizing the Imagination: Disney's Wilderness Lodge. *Capitalism, Nature, Socialism* 8: 107–130.

Donlan, J., et al. 2006. Pleistocene Rewilding: An Optimistic Agenda for Twenty-first Century Conservation. *American Naturalist* 168: 160–183.

Egan, D., and E. Howell. 2001. *The Historical Ecology Handbook: A Restorationist's Guide to Reference Ecosystems.* Washington, DC: Island Press.

Hall, M. 2005. *Earth Repair: A Transatlantic History of Environmental Restoration.* Charlottesville: University of Virginia Press.

Hall, M. 2010. *Restoration and History: The Search for a Usable Environmental Past.* New York: Routledge.

Hallett, D., and L. Hills. 2006. Holocene Vegetation Dynamics, Fire History, Lake Level and Climate Change in the Kootenay Valley, Southeastern British Columbia, Canada. *Journal of Paleolimnology* 35: 351–371.

Harris, J., R. Hobbs, E. Higgs, and J. Aronson. 2006. Ecological Restoration and Global Climate Change. *Restoration Ecology* 14: 170–176.

Higgs, E. 1997. What Is Good Ecological Restoration? *Conservation Biology* 11: 338–348.

Higgs, E. 2003. *Nature By Design: People, Natural Process, and Ecological Restoration.* Cambridge, MA: MIT Press.

Higgs, E. 2007. Twinning Reality, or How Taking History Seriously Changes How We Understand Ecological Restoration in Jasper National Park. In *Culturing Wilderness in Jasper National Park Studies in Two Centuries of Human History in the Upper Athabasca River Watershed,* ed. I. MacLaren, 289–316. Edmonton, Canada: University of Alberta Press.

Hobbs, R., and V. Cramer. 2008. Restoration Ecology: Interventionist Approaches for Restoring and Maintaining Ecosystem Function in the Face of Rapid Environmental Change. *Annual Review of Environment and Resources* 33: 39–61.

Hobbs, R., L. Hallett, P. Ehrlich, and H. Mooney. 2011. Intervention Ecology: Applying Ecological Science in the Twenty-first Century. *Bioscience* 61: 442–450.

Hobbs, R., E. Higgs, and J. Harris. 2009. Novel Ecosystems: Implications for Conservation and Restoration. *Trends in Ecology & Evolution* 24: 599–605.

IPCC. 2007. *Climate Change 2007: Synthesis Report,* Contribution of Working Groups I, II and III to the Fourth Assessment Report of the Intergovernmental Panel on Climate Change, ed. R. K. Pachauri and A. Reisinger. Geneva, Switzerland: IPCC.

Jackson, S., and R. Hobbs. 2009. Ecological Restoration in the Light of Ecological History. *Science* 325: 567–568.

Levesque, L. 2005. "An Integrated Landscape Change and Ecological Restoration: An Integrated Approach Using Historical Ecology and GIS in Waterton Lakes National Park, Alberta." M.Sc. thesis, University of Victoria.

Lowenthal, D. 2010. Reflections on Humpty-Dumpty Ecology. In *Restoration and History: The Search for a Usable Environmental Past,* ed. M. Hall, 13–34. New York: Routledge.

Luckman, B. H., and T. A. Kavanagh. 2000. Impact of Climate Fluctuations on Mountain Environments in the Canadian Rockies. *Ambio* 29: 371–380.

MacDougall, A., B. Beckwith, and C. Maslovat. 2003. Defining Conservation Strategies with Historical Perspectives: A Case Study from a Degraded Oak Grassland Ecosystem. *Conservation Biology* 18: 455–465.

MacLaren, I. 1999. Cultured Wilderness in Jasper National Park. *Journal of Canadian Studies [Revue d'Etudes Canadiennes].* 34 (3): 7–58.

MacLaren, I., E. Higgs, and G. Zezulka-Mailloux. 2005. *Mapper of Mountains: M.P. Bridgland in the Canadian Rockies 1902–1930.* Edmonton, Canada: University of Alberta Press.

Meine, C. 1991. *Aldo Leopold: His Life and Work.* Madison: University of Wisconsin Press.

Parks Canada. 2008. *Principles and Guidelines for Ecological Restoration in Canada's Protected Natural Areas.* http://www.pc.gc.ca/docs/pc/guide/resteco/index_e.asp (accessed September 6, 2011).

Rhemtulla, J., R. Hall, E. Higgs, and E. MacDonald. 2002. Eighty Years of Change: Vegetation in the Montane Ecoregion of Jasper National Park, Alberta, Canada. *Canadian Journal of Forest Research* 32: 2010–2021.

Rosenzweig, M. 2003. Reconciliation Ecology and the Future of Species Diversity. *Oryx* 37: 194–205.

Roush, W. 2010. "A substantial upward shift of the alpine treeline ecotone in the southern Canadian Rocky Mountains." M.Sc. thesis. University of Victoria, BC.

Society for Ecological Restoration International. 2004. "The SER International Primer on Ecological Restoration," Science & Policy Working Group. http://www.ser.org (accessed November 15, 2011).

Suding, K., and K. Gross. 2006. The Dynamic Nature of Ecological Systems: Multiple States and Restoration Trajectories. In *Foundations of Restoration Ecology*, ed. D. Falk, M. Palmer, and J. Zedler, 190–209. Washington, DC: Island Press.

Tucker, N. 2005. Healing Country and Healing Relationships. *Ecological Management & Restoration* 6 (2): 83–84.

United States National Park Service. 2010. Antietam National Battlefield, Maryland. http://www.nps.gov/anti/index.htm (accessed February 2, 2010).

Vellend, M., A. Bjorkman, and A. McKonchie. 2008. Environmentally Biased Fragmentation of Oak Savanna Habitat on Southeastern Vancouver Island, Canada. *Biological Conservation* 141: 2576–2584.

White, P. S., and J. L. Walker. 1997. Approximating Nature's Variation: Selecting and Using Reference Information in Restoration Ecology. *Restoration Ecology* 5 (4): 338–349.

II

Integrating Ecology into the Virtue of Justice

5

The Death of Restoration?

Andrew Light

The science and practice of restoration ecology has traditionally been tied to some understanding of environmental history. As opposed to other kinds of environmental management, the justification primarily used for an ecological restoration is to return an ecosystem, a place, or even a species to some state representative of some point in its past in order to glean some environmental benefit such as mitigating pollution or maintaining and increasing biodiversity.[1] While invoking history as a parameter for selecting and designing restoration projects has sometimes led to problems—most notably the unfortunate period in North America when many labored under the assumption that the proper goal of restoration should be to restore to pre-Columbian conditions because this was a more "natural" state—in most respects the historical focus of restoration has served its advocates well insofar as it provides a set of reasons to distinguish restoration from other forms of environmental management or intervention.

Different restorationists orient themselves around "historical fidelity" to different extents and degrees often depending on the subject of their work. While terrestrial restorationists, such as those focusing on tallgrass prairies and forests, often seem most interested in returning a place to the same landscape that actually existed in that place before (which I will later call a "token restoration"), those focusing on riparian restoration generally set aside historical fidelity given the extreme difficulty of returning streams and rivers to their original or "natural" flows given changes in flood control and water delivery over time (Seavy et al. 2009).

Nonetheless, what all restorationists seem to agree on now is that current and future climatic change caused by anthropogenic global warming will challenge their practice. For some this only means that reproducing a historical reference condition will no longer be as high a priority as it might have been before and that other criteria for success will supersede. For others the certainty of a shifting climate creates an existential crisis of sorts. As average temperatures change at a given latitude the ecosystems and ecological communities that were historically present in a particular place may no longer be viable or resilient there now or in the future. One cannot restore a token of an ecosystem—which was once in a particular place before it was changed by humans—when that ecosystem is no longer adaptable to the new climate that is expected to predominate in the same place. To do so would be pointless from an ecological point of view, and so restoration becomes practically impossible if the overall goal of the project is tied to any long-term ecological benefit. Thomas Simpson, an ecologist at the Glacial Park Research Field Station in Illinois, puts the relationship between historical fidelity as a goal of restoration and the emergence of a new global climate this way: "Ecosystems are not abstractions, a sum of living things and processes; they are part of a place, and to restore the ecosystem is to restore the place to a semblance of its former condition. What I or any resource manager can *restore* in northeastern Illinois is limited to what came before" (2009, 343). One impact of climate change could be to make this kind of restoration, in his words, "impossible."

I will not rehearse the arguments here that humans are warming the planet and that this will have long-term and possibly irreversible impacts on global ecosystems (I recommend Lovejoy and Hannah's masterful and comprehensive *Climate Change and Biodiversity* [2005]). Even if all greenhouse gas emissions on the planet stopped today there is enough carbon in the global system to ensure some warming over pre-industrial levels and therefore some changes to ecosystems around the globe. Of course, these emissions will not stop today and prospects of limiting projected temperature increases to two degrees Celsius over pre-industrial levels—which has now been accepted by most as the best achievable outcome of mitigation efforts—are uncertain at the very best.

Restorationists are therefore faced with twin moving targets. On the one hand, even while regional and local environmental impacts remain one of the most uncertain areas of climate science, the species and systems that restorationists want to restore are literally on the move, usually to cooler places where they may not have historically been established before. On the other hand, what defines restoration ecology and distinguishes it from other forms of environmental management is also moving and as a result fundamentally challenging many of the core principles of the field. As Harris, Hobbs, Higgs, and Aronson put it in their signature statement on the impact of climate change on restoration:

"The critical question facing us is to elaborate appropriate strategies and tactics for restoration as thus defined in a world of rapidly changing climate regimes, when in many cases relying on historical references makes less sense" (Harris et al. 2006, 172).

Does climate change then imply the death of restoration? Will it give way to a new focus on the creation of wholly new engineered ecosystems to adapt to changing environmental conditions? I will argue that climate change may not necessarily end restoration so long as restorationists, ecologists, and environmentalists continue to disentangle restoration from narrow expectations of historical fidelity that do not reflect the current state of restoration practice. Further, the challenge of the moving targets may present an opportunity to once again emphasize the importance of social and cultural goals in restoration practice as guiding criteria for selection of restoration projects.

Defining Restoration: History in Practice

Two key questions confront us when we examine the relationship between restoration ecology and climate change:

1. How will climate change affect the practice of restoration ecology?

2. How can restoration ecology assist in our response to climate change?

I will focus much more attention on question 1 in what follows for the obvious reason that if the answer to (1) is that climate change spells the death of restoration as a practice, then the answer to question 2 will be "very little" and those who have been focusing on restoration as a key component of productive environmental management, stewardship, or citizenship should turn to other pursuits if they are interested in using their talents to respond to climate change.[2]

One answer to (1) that avoids this predicament can be found by focusing on the definition of "ecological restoration" as it reflects actual practice in the field. If the definition of "restoration" can be decoupled from its traditional reliance on historical reference points and still be distinctive from other forms of environmental management, then the impact of climate change on ecological restoration may only be to diminish historical fidelity to a criteria of project evaluation and not necessarily challenge the essence of the practice.

How is restoration defined? Let's start with the primary organization for restoration ecologists: the Society for Ecological Restoration International (SERI). Over the years it has tried to define restoration. At first glance, three examples of these exercises reveal a continued reliance on history as a definitive component of the practice.

1990   "Ecological restoration is the process of intentionally altering a site to establish a defined, *indigenous, historic* ecosystem. The goal of this process is to emulate the structure, function, diversity and dynamics of the specified ecosystem." . . .

1996   "Ecological restoration is the process of assisting the recovery and management of ecological integrity. Ecological integrity includes a critical range of variability in biodiversity, ecological processes and structures, regional and *historical* context, and sustainable cultural practices." . . .

2002   "Ecological restoration is the process of *assisting the recovery* of an ecosystem that has been degraded, damaged, or destroyed." (Higgs 2003, 107–110; emphasis added)

The first thing to note is that explicit reference to historical conditions drops out by the last definition. Nonetheless history is still strongly implied in this definition. If ecological restorations assist in the recovery of an ecosystem, then presumably one measure of recovery is whether an ecosystem returns to its predisturbance condition.

While these definitions provide evidence of a reliance on history, we should not stop here. How this professional society defines restoration by impaneling a committee to do so does not necessarily tell us what restoration really is or whether achieving a historical benchmark is really necessary to the practice. While I will not attempt here a comprehensive overview of the attempts to distill a more thoroughly defended definition of restoration in the literature, a representative example should suffice in demonstrating the difficulty of prising out historical conditions from the definition of restoration. I'll focus on Eric Higgs both because he has produced what I find to be the most ecologically informed treatment of the philosophical dimensions of restoration so far and because he was a witness to and participant in the various institutional attempts to define restoration in the SERI (Higgs 2003). We can see in Higgs's treatment of the subject evidence of significant reliance on history as a criteria for distinguishing restoration from other environmental practices.

Higgs defines ecological restoration as a form of environmental management characterized by attention to historical fidelity and ecological integrity (2003, 95). By historical fidelity he means "loyalty to predisturbance conditions" (or at least minimally that projects have a "historically motivated goal"). By ecological integrity Higgs defends three criteria: (1) structural or compositional replication of a reference ecosystem, (2) functional success, and (3) durability. Higgs also goes a step further and uses the definition as the basis for an claim about evaluative criteria: A good restoration on his account is one that is characterized by attention both to historical fidelity to predisturbance conditions as well as recreation of the ecological integrity of a site.

Now before proceeding further, it should be noted that the scope of historical fidelity in Higgs's account is not only ecological, but cultural as well. In *Nature by Design* he argues that the mistake of many definitions of restoration is that they focus too much on technical proficiency and do not provide an "indication of the wider cultural context of restoration practice" (2003, 108). Good restorations on this account should pay attention to natural and cultural elements insofar as we can meaningfully distinguish between those elements at any particular site. While I think that we need to understand the social and cultural dimensions of restoration (indeed, I will argue at the end of this chapter that they are a key part of preserving and advancing the practice of restoration in a changing climate), I have argued elsewhere that we ought to try to keep apart descriptive and normative accounts of restoration (Light 2009). My concern is that mixing them lends itself to a tendency to derive our moral and social assessment of restorations from our description of the kind of thing we take them to be and then in turn to narrow what counts as a restoration to those projects which most effectively exhibit these predetermined criteria.

But my core point here is not to take apart Higgs's criteria for what counts as a restoration or a good restoration but rather to examine what they can tell us about the limitations of existing definitions of restoration, especially in terms of their focus on historical criteria. For when we look at actual examples of restorations, what we encounter is a huge variety of projects, some of which only bear a tangential relationship to historical reference states, even though most would admit to their value as restorations. The fact that any given project, or cluster of projects, do or do not exhibit different properties does not mean they are not restorations. They just may not be good restorations under some particular description.

Let's briefly look at four examples of restorations in the United States, the first three of which are well known in the restoration literature. We will see that however we want to define what restoration is, as a practice it has evolved into a plethora of projects which are strikingly different from one another on Higgs's definition of restoration as well as on most others.[3]

Curtis Prairie, University of Wisconsin Arboretum
Originating in the 1930s, the sixty-acre Curtis Prairie is often described as the world's oldest restored tallgrass prairie, featuring big bluestem and Indian grasses and reestablishing a prairie that had existed on the site as late as 1835. It was a mainstay of the establishment of the early science of restoration, including experiments with controlled burning regimes to stimulate new growth and prevent successions of trees and shrubs that choke out prairie plants.

Comprehensive Everglades Restoration Plan (CERP)
Developed in the 1990s and authorized by an act of the U.S. Congress in 2000, CERP is a $10.5 billion thirty-year project (with more than fifty individual components) encompassing over four hundred thousand acres. It is designed to enhance Everglades wetlands and associated lakes, rivers and bays in the sixteen-county region of Florida and capture and store water to supply the booming population of the area's cities.

Chicago Wilderness
Embedded in one of the largest metropolitan regions in the United States, the Chicago Wilderness is a network of public and private protected natural areas including 370,000 acres of protected land and waters spanning over four states. While the entire network is not a giant restoration project like CERP, the original core of the Chicago Wilderness consisted primarily of restorations of oak savannah in the Chicago forest preserves—thin strands of original forest which once ringed the outskirts of the city but are now embedded inside it (see Stevens 1995). These restorations require constant maintenance to ensure that they do not succumb to succession regimes or to the encroachment of exotic and ornamental species from the surrounding urban and suburban landscapes.

Vintondale Acid Mine Drainage (AMD) Project
Started in 1994 by historian T. Allan Comp, the Vintondale AMD project aimed to clean up toxic acidic water from an abandoned mine complex in Southwestern Pennsylvania, and included restoration of a wetland designed to recharge the water after it had been passively cleaned of contaminants. Rather than only restoring the wetland and integrating it into a larger pollution mitigation project, the restorationists also attempted a thoroughgoing "holistic" renewal of the abandoned mine complex that highlights the old industrial legacy rather than removing it from the landscape. Visitors not only learn about the problem of acid

mine drainage which plagues the Appalachian region and how to clean it up, but also the legacy of the mining community in Vintondale through vestiges of the preserved industrial legacy of the place—such as the intact foundations of the coal washing facility on the site which border and effectively house the wetland.

When we compare these projects on Higgs's criteria we get some interesting results.

Project	Historical fidelity
Curtis Prairie	Yes
Everglades	Some/No
Chicago Wilderness	Yes
Vintondale	Some/No

Do these four projects evince a "loyalty to pre-disturbance conditions," as Higgs puts it? Certainly the Curtis Prairie does what it was designed for, specifically to replicate the prairies in that area, even on that site in the past. It is a prime example of a "token restoration" insofar as it tries to recreate the ecosystem that existed there before. In addition to trying to discover and replicate the mosaic of plants that made up tallgrass prairies, restorationists working on the Curtis also physically replanted sod from remnant patches of prairie in the area.

So too with the restorations in the Chicago Wilderness, at least the recreation of what the restorationists argue were the oak savannahs that once dominated the Chicago forest preserves. But unlike the prairies in Wisconsin, there has been some controversy over the existence of this kind of ecosystem in this region in the past. In this respect, the Chicago restorationists could be restoring a "token" ecosystem of one that actually was here in the past or rather this could instead be a "type restoration," or a restoration of a type of ecosystem that may have existed in a place relatively similar to this one in the past. (Some argue that type restorations are less preferable than token restorations in terms of adherence to historical fidelity.[4]) Whether this is a token or type restoration does not necessarily matter here in terms of evaluating the historical fidelity of the project. Restorationists working on this project, such as Steve Packard, have long defended their choices on the basis of ecological and historical arguments that what they are doing is loyal to the reference ecosystem that they claim to have discovered. They employ this claim to justify their removal of the existing dense forest (dominated by exotic European buckthorn and other leafy trees) to create a more open canopy of oaks allowing for a semi-shaded prairie ecosystem.

But when we look closer at the Everglades restoration plan and the Vintondale AMD project, while we find some attention to historical fidelity, it is certainly not as thoroughgoing as the previous cases. CERP has been widely criticized for a variety of reasons. Nonetheless, our initial evaluation of its adherence to historical fidelity should—as we saw with the Chicago Wilderness—come through an understanding of the intentions behind the project. While there are many reasons for the Everglades restoration and many ecological benefits that it could potentially produce, my own view is that the core argument that made it politically possible was the need to provide a reliable source of clean drinking water in southern Florida. More than anything else, this human need drove the bipartisan consensus that released both federal and state funds to pay for the array of projects that will go into the restorations.

Setting aside for now the parts of the project which will reestablish critical habitat for endangered species, those components aimed at reversing decades of agricultural impacts on the ability of the wetlands to filter and recharge water will recreate riparian systems which of necessity must conform to the new land-use patterns that have emerged in the area and so cannot hope to aspire to achieving a token restoration. So, as engineers and restorationists reverse the channelization of rivers in the system, such as the Kissimmee River, which was straightened out by the Army Corps of Engineers in the late 1960s as a method of flood control, they will try to reverse a process which dried out some forty thousand acres of floodplain and which had detrimental impacts on human and nonhuman communities. The current plan calls for restoring forty-three miles of the Kissimmee, though it cannot as a matter of course replicate the original flow of the river as the resulting floodplain would threaten human communities in the area. Moreover, if the Kissimmee had never been channelized, it would be different today—thus requiring a different kind of restoration plan than one that attempted to recreate the original flow of the system. At best, most of the restorations that are part of the overall CERP will be type restorations, though very broad types at that.

Similarly, the Vintondale AMD project does not aim to replicate the historical landscape or ecosystems that surrounded the small mining town of Vintondale. While there were wetlands that were destroyed by the mining operation on the site, the wetland being created is not a token replica of one that was there before. It is oriented so that it receives water from a filtering facility that increases the pH of the water coming out of the mine. More important, whatever original wetlands there were in the area were certainly not banked by the crumbling remains of the foundation of an abandoned part of the Vinton Colliery. However, by cleaning up the contaminated water emerging from the abandoned mines in the area, the restorationists in Vintondale are attempting to restore the ecosystem health of Blacklick Creek, which flows through the site taking in water from the restored wetland.

As to Higgs's second criteria, "ecological integrity," we have three parts of these projects that could be evaluated: structural composition, functional success, and durability. Briefly, all four projects seem to evince success at structural composition, or at least try to with reference, minimally, to a type of ecosystem which was predominant in the area. I will turn in a moment to the question of functional success. What I want to focus on now is the last part of the integrity criteria: durability.

Durability is an interesting and troubling requirement or goal for restoration, since it implies some measure of persistence over time for systems that really ought not persist over time due to natural succession. Many have argued that the real measure of durability is whether a restored system can "maintain itself" over time (though what that really means for any given ecosystem is often in dispute). Critics have argued that one of the things that make a restoration not really "natural" is that it cannot persist without human assistance (discussed in Light 2009). Of course, some ecosystems that have not been altered by humans remain relatively stable and others move on to other successional states. In turn, some restorations seek to restart natural processes and let them evolve accordingly—especially minimal restorations which work essentially by removing a contaminant from a system and then letting it proceed on its own course—while others try to artificially replicate parts of a system which create relative stability that withstands succession (see Throop, chapter 2, this volume).

Prairie restorations are a fascinating example of the latter. Tallgrass prairies, and systems which contain elements of prairie such as oak savannahs, must replicate the role that fire plays in the creation and maintenance of these systems. Prairie ecosystems evolved with a regime of regular natural burning that not only augments reproduction of some flora species in the prairie mosaic but also keeps back the growth of woody species that would shade out grasses. But usually for reasons of scale—the difficulty in restoring a prairie large enough that it would naturally burn from lightning strikes—or because of the dangers of the proximity of restored systems to property (which could be destroyed by fire), prairie and savannah restorationists have to use carefully prescribed burns to maintain these systems.

Since two of our preceding examples involve this kind of management, again we see mixed results in an evaluation of this part of Higgs's criteria.

Project	Historical fidelity	Integrity: Durability
Curtis Prairie	Yes	Possible/No
Everglades	Some/No	Yes
Chicago Wilderness	Yes	Possible/No
Vintondale	Some/No	Yes

Both the Curtis Prairie and the Oak Savannahs in Chicago require active management that intervenes by stopping succession through controlled burning. If "durability" in restoration ecology means creation of a system that maintains relative stability on its own over time, or allows for natural succession, then neither project is a restoration or a good example of a restoration by this criterion. However, one could argue that this is an unfair criterion for evaluation of all projects. The characteristics of the kind of ecosystem that one is trying to restore should count for something and this kind of ecosystem is just not amenable to durability under that description. If by "durability" we instead mean something more like assisted stability over time—which would be reasonable since many restorations fail even with a lot of continued help—then these two might count as fulfilling that part of ecological integrity. As for the Everglades restorations and the Vintondale project, both riparian systems are designed for durability, as we understand it to mean unassisted stability, though it must be recognized that both will require continuous monitoring to ensure that various human-made parts of these systems involving pollution regulation do not fail.

My conclusion from this exercise is that neither historical fidelity nor ecological integrity overall is satisfactory as a basis for determining whether a project is a restoration, and likely whether it is a good restoration, when we take these definitions out into the field. This is good news for resolving the worry that the traditional focus in defining restoration—as a practice concerned with recreating systems based on historical reference conditions—spells its demise in an age of global warming.

More important, working through these four cases helps us to see that context matters both in terms of defining and evaluating the success of restoration projects. Higgs does a good job of carefully defending each part of his criteria as a part of what should count as a restoration and even a good one. But unless one is prepared to claim that not meeting some part of this criteria means that a given project is not a restoration—or necessarily is a bad restoration—then historical fidelity as it is traditionally understood is simply not the *sine qua non* of restoration (this conclusion can be compared with Sandler, chapter 3, this volume). The essence of a restoration is defined in large part by the purpose and constraints of the project. It is no surprise that the two preceding examples that primarily involve riparian areas may fail the definition of historical fidelity, given that adherence to it is not necessarily conducive to creating a functional stream or river. Nor is it a surprise that the two examples that are sketchy on durability are fire-dependent grasslands. Members of the restoration community do not generally write off these projects as something other than restoration because, as adherents of good ecological science, they understand that these differences matter. So too when evaluating the success of these projects, the context and ecological goals of a restoration count more than adherence to an abstraction such as historical fidelity. It would be absurd to say that the best riparian restorations are always worse than the best forest restorations because of the limitations of historical reference conditions as a guide to riparian restoration. We ought to evaluate like cases alike.

In this respect, whether a restoration is constrained by future changes in the climate should not count as a reason to discount a project as a restoration. If attempting a token restoration in one place is unadvisable because changes in the climate will make it unstable, then we ought not to try it. But this does not mean that all restoration activities in places vulnerable to climatic change are second best.

Before leaving this section, though, a brief word is in order concerning a core definition of restoration that I think is workable, regardless of concerns about climate change. All of the projects previously mentioned either do meet or clearly aspire to meet the second part of Higgs's definition of ecological integrity: functional success. They all aim to restore or recreate a function of an ecosystem or ecological process relative to its current state of disturbance and then take appropriate measures to ensure that functional success.

Interesting here as well is that if we go back and look at the SERI definitions, they also employ the language of functions. The 1990 definition says this explicitly: "The goal of this process is to emulate the structure, *function*, diversity and dynamics of the specified ecosystem." The 1996 and 2002 definitions imply it by appealing to the creation of something that actively does something that assists recovery of some state or process. The underlying intuition is that something is a restoration when it is an attempt to restore or recreate the function of a previously existing ecosystem, a component of that ecosystem, or an ecosystem service provided by a reference ecosystem (such as habitat for endangered species, recharge of a water supply, and so on).[5]

For similar reasons I have proposed the following definition: *Ecological restoration is a form of environmental intervention which attempts to recreate some aspect of the prior function of an ecological reference state* (Light 2009, 153). I discuss more of the philosophical issues involving this definition in the paper where this definition is introduced. Here only note that accepting this definition does not depend on one's answer to the issue of whether all ecosystems have functions or whether all aspects of all ecosystems can be described as a function of that ecosystem. Rather, what matters is whether we can attribute functions to ecosystems that could be replicated for some reason. As such, the definition is agnostic on some of the stickier issues concerning biological functions. Restorationists are not trying to produce biological entities as such, but rather are making something that reproduces functional attributes of specific kinds of systems in nature. To plan our restorations on our understanding of these attributes does not mean that all aspects of a restored site can be reduced to the recreation of functional properties. Designs for restorations can include other elements, such as aesthetic components, even while our definition of the practice of restoration is first found in these functional properties.

But for the purpose of the present discussion, a successful defense of this definition would solve both of the moving-target problems mentioned at the start. First, if we fully exploit the final "an" in the definition, then there should be no problem in abandoning the goal of prioritizing token over type restorations and so we won't be necessarily compelled to move the site of a restoration to conform to a changing climate. So, too, with the moving target of what restoration means. In a changing climate, different aspects of restoration projects will become more important. Harris, Hobbs, Higgs, and Aronson put the point this way:

The principal objects of manipulation are the abiotic and biotic components [of a system], which we seek to change to overcome abiotic and biotic barriers. For example, we may need to restore a dynamic water level regime, in terms of both amplitude and frequency, or reinstate heterogeneity to convert arable land to forest; these are both abiotic interventions. We may need to reintroduce species at a variety of trophic levels, from soil symbionts, through primary producers to herbivores and predators—some of which were not recently found on this site. Hence, an increasing emphasis will be on proper functioning condition of a site—ecological integrity—and to a lesser extent on nudging a site back to historical conditions based on species. In general, process, not structure, will prevail. (Harris et al. 2006, 173–174)

This does not mean, contra Ronald Sandler's argument (chapter 3, this volume), that one impact of global warming on restoration is to make it less important going forward. To take a cue from the 2002 SERI definition, which is otherwise quite vague, a perfectly reasonable and respectable goal of any restoration project is to assist the recovery of a degraded ecosystem not necessarily to some previous state but to reflect the best function that is possible for it under changing climatic conditions. For those interested in place-based restorations (by which I mean reversing the history of degradation in a particular site), they should no longer feel that they could only succeed by returning that place to a prior state. For those more concerned with preservation of a specific species or cluster of integrated species (such as we find in chapter 6 by Keulartz and Swart in this volume), then they may need to find the new places where those species could potentially flourish in a climatically altered world. There will be debates and disputes over what kind of restoration we should try to do in any given place given competing priorities, but this has always been the case for any restoration project before the full implications of climate change were appreciated by the restoration community.

History need not be abandoned writ large in restoration, only the simplistic ideal that the best restoration is that which recreates some historical snapshot in the history of a place. Instead, we could understand history as playing three roles in any restoration. First, if it is possible, then a restoration should try to restore a vestige of the functional system that was there before, especially if it is disappearing elsewhere. That said, no restoration is necessarily better because the relative climatic stability of its environment is able to reflect a historically accurate reference system. Second, following Higgs's contribution in this volume, restorations should always aim to maximize open-ended ecological possibilities insofar as those possibilities can be judged to promote overall ecological integrity, health, or welfare (see Holland, chapter 7, this volume, on this point as well). One measure of those possibilities can be historical insofar as there are advantages to our ability to identify a compelling narrative about the continuity from the past to the future in any place (there is much more that can be said on this point, though interested readers should consult the third section of O'Neill, Holland, and Light 2008).

And third, what we should be trying to restore is not necessarily only "nature" but also the human relationship with nature that, as we saw with the Vintondale AMD project, clearly has a complicated historical component. More on this point follows.

Avoiding Hubris

If the reality of anthropogenic climate change will not necessarily destroy the coherence of restoration, then it could, to take up the original second question raised early in this chapter, assist in our response to climate change. Several authors have already articulated different ways in which restoration can and already does contribute to our struggle with global warming. For example, Harris and company argue that it is essential to preserve vestiges of species assemblages important for biodiversity which would otherwise be lost (Harris et al. 2006, 174). Seavy and colleagues argue that riparian restoration specifically is an absolutely necessary component of promoting resilience of ecosystems to climate change (Seavy et al. 2009, 331), and Lovejoy and Hannah offer a forceful presentation of the claim that mitigation efforts cannot hope to approach critical mass without intensive use of restoration (2005, 387–395). Given that global deforestation accounts for annual CO_2 emissions equal to the entire global transportation sector, and given the essential role of improved land use and expanded forests as a carbon sink, a huge amount of carbon mitigation will involve restoration and preservation of forests. We can no more hope to achieve the two-degree-Celsius target without forest restoration than we could without moving to alternative transportation fuels.

Nonetheless, some will be concerned that backing away from the guiding role of history in identifying and evaluating restorations will lead to a slippery slope where anything goes. Critics of restoration such as Eric Katz have long been concerned (as can also be seen in many of the chapters in this volume) that restoration already promotes a form of hubris. We will be less concerned about doing harm to nature if we think we can easily fix it. While I have long argued that such concerns are greatly exaggerated (see Light 2000), a slightly different problem with hubris haunts the present discussion. It is perhaps best voiced in this volume by William Throop's concern that a diminished reliance on history will open up the Pandora's box of an increased focus on engineered ecosystems limited only by human aspiration. If the goals of restoration are increasingly untethered from historical reference conditions then there may be nothing to stop us from abandoning any pretense to not allowing human desires and preferences govern our decisions about the future of different places.

While I hope I have provided a useful role for history in restoration at the end of the last section the fact of the matter is that very little could stop anyone from using restoration, or any other ecological management tool for that matter, to justify something which we may find either bad for other species or ecosystems or bad for setting a precedent for how we should live with nature in our emerging warmer world. One could try to form a governing ethos to guide restoration as a whole, articulate and defend a philosophy of restoration, or, as Ned Hettinger puts it in chapter 1 of this volume, promote a "paradigm" for restoration to limit its use and abuse. But as Hettinger demonstrates, such attempts are more trouble than they are worth and beg far too many questions. Restoration ecology is a management tool that can be used for better and worse purposes. (I think though that most of the worse purposes uncovered in the literature are merely misguided decisions made with imperfect information.) Constructing a grandiose paradigm for restoration strikes me as just another attempt to stipulate what a good restoration is rather than coming up with defensible evaluation criteria appropriate to the context of different kinds of restoration.

Rather than a paradigm for restoration, I have defended the claim that what restorations should try to achieve should go beyond a set of ecological goals. Each of these goals needs to be defended independently. So, because I have argued that restoration (along with other activities) is highly appropriate for engaging nonexperts in important environmental practices, then—all other things being equal—restorations should try to maximize opportunities for voluntary participation. This is because they can potentially help to create a kind of ecological citizenship between people and the places around them (Light 2000, 2002). Since the vast majority of the labor in the Chicago forestry preserves was completed by volunteers, that process provides an expanded justification for those projects. In a forthcoming work, I will defend the idea that restorations should, where possible, preserve the "disturbance memory" of a site— leaving some vestige of the means that caused the disturbance to the system in the first place. The Vintondale AMD project is a particularly good example of capturing that value which in turn offsets some of the concerns others may have about the project.

Again, these various criteria do not ensure that some restorationists will not imagine themselves to be unencumbered masters of the planet but, insofar as these criteria might embody positive virtues of the human relationship with nature, they will raise the bar on what counts as the best kind of restoration. And as we identify more of the virtues necessary for humanity in a warming world (which the editors and many of the authors of this volume endeavor to do) to the extent that those virtues can be defended, then forms of restoration that embody them will be preferable.

Restoration will neither damn us nor save us in the years to come but it will be part of a cluster of practices which could represent the best that we can do in an uncertain future.

Notes

1. I use "restoration ecology" and "ecological restoration" interchangeably. There have been numerous attempts over the years to carve out a discrete definition or realm for each—such as the claim that "restoration ecology" should refer to the science and "ecological restoration" should include the cultural and social aspects of the practice of restoration—but I have always found these distinctions artificial and unhelpful. I do not believe that any of my arguments here stand or fall on accepting or denying such distinctions.

2. There are still active debates in the literature over whether the human relationship with nature should be one governed by an ethos of management, stewardship, citizenship, cohabitant, or some other framework. Those debates do not concern me here, and I do not think they matter for my arguments in this chapter. I have previously argued that citizenship is the preferred framework for understanding the full moral and social potential of restoration (see Light 2002).

3. I encourage readers to look up and briefly examine the websites for these four projects, which are comprehensive and contain excellent photographs that will make the following discussion more palpable.

4. This token-type distinction originates with the work of Robert Elliot (1997).

5. Compare Light's remarks on a plurality of ecosystem services as providing legitimate goals for ecological restoration with Holland's ideas about the environment as a meta-capability, necessary for realizing other more basic human capabilities (chapter 7, this volume), and with Schlosberg's ideas about the capabilities of an ecosystem itself (chapter 8, this volume). Under Light's pluralistic view, enabling various capabilities may also provide legitimate restoration goals.—eds.

References

Elliot, Robert. 1997. *Faking Nature*. London: Routledge.

Harris, James A., Richard J. Hobbs, Eric Higgs, and James Aronson. 2006. Ecological Restoration and Global Climate Change. *Restoration Ecology* 14: 170–176.

Higgs, Eric. 2003. *Nature by Design*. Cambridge, MA: The MIT Press.

Light, Andrew. 2000. Ecological Restoration and the Culture of Nature: A Perspective, Pragmatic. In *Restoring Nature*, ed. P. Gobster and B. Hull, 49–70. Washington, DC: Island Press.

Light, Andrew. 2002. Restoring Ecological Citizenship. In *Democracy and the Claims of Nature*, ed. B. Minteer and P. Taylor, 153–172. Lanham, MD: Rowman and Littlefield.

Light, Andrew. 2009. Ecological Restoration: From Functional Descriptions to Normative Prescriptions. In *Functions in Biological and Artificial Worlds*, ed. P. Kroes and U. Krohs, 147–162. Cambridge, MA: MIT Press.

Lovejoy, Thomas E., and Lee Hannah, eds. 2005. *Climate Change and Biodiversity*. New Haven, CT: Yale University Press.

O'Neill, John, Alan Holland, and Andrew Light. 2008. *Environmental Values*. London: Routledge Press.

Seavy, Nathaniel E., et al. 2009. Why Climate Change Makes Riparian Restoration More Important than Ever. *Ecological Restoration* 27: 330–338.

Simpson, Thomas B. 2009. Old Nature, New Nature: Global Warming and Restoration. *Ecological Restoration* 27: 343–344.

Stevens, William K. 1995. *Miracle Under the Oaks*. New York: Pocket Books.

6

Animal Flourishing and Capabilities in an Era of Global Change

Jozef Keulartz and Jac. A. A. Swart

In part I of this book a core assumption of ecological restoration is seriously being challenged: are historical baselines or benchmarks still of any use for setting goals in ecological restoration in an era where human impact on the environment is omnipresent to such an extent that Nobel Prize winner Paul Crutzen has introduced the term "Anthropocene" (cf. Thompson and Bendik-Keymer, introduction to this volume)?[1] Can historical fidelity still be considered an appropriate value or virtue during such an era of large-scale human-induced environmental change? In the face of this unprecedented change, restoration to a historic standard seems to become more and more anachronistic: "Valuing the past when the past is not an accurate indicator for the future may fulfill a nostalgic need but may ultimately be counterproductive in terms of achieving realistic and lasting restoration outcomes" (Harris et al. 2006, 175). Ronald Sandler (chapter 3, this volume) seems to agree with this statement and argues that the ecological justification for historical fidelity is weakened or diminished, whereas both William Throop and Eric Higgs (chapters 2 and 4, respectively, this volume) favor a continued critical role for historicity in restoration.

In this chapter, we want to go into another threat that ecological restoration projects are facing during the emerging Anthropocene, namely the threat to animal welfare and flourishing. Today, the animal world is under severe attack as a result of two strongly interconnected global processes. On the one hand, globalization causes massive dislocations of entire populations. As trade, travel, transport, and tourism boom, the world is becoming more and more borderless and, by the same token, it is becoming increasingly vulnerable to invasive species. Since globalization took off, more plants and animals have become globetrotters than ever before. We are being confronted with the problem of how to control invasive species that threaten to degrade or destroy ecosystems.

On the other hand, global environmental changes such as climate change, land-use and land-cover change, and deforestation and desertification are having disruptive impact on plant and animal life. Entire populations are being confronted with the alternative to abandon their original habitat or become extinct. As a result of the rapid fragmentation of natural habitats, nature areas are being turned into "islands in a sea of cultivated land." We are witnessing the ongoing conversion of what were originally continuous populations to "metapopulations," which are spread geographically over the dwindling patches of suitable habitat that remain. Because these patches are usually small and can only accommodate small populations, and because the movement of the animals between these patches is restricted for lack of connectivity, local extinction of subpopulations is a common event. This situation asks for metapopulation management, by which one may try to artificially perform the function of dispersal and recolonization of patches of locally endangered or extinct species.

With metapopulation management the distinction between classic in situ and ex situ work is gradually breaking down. As the size and genetic diversity of remaining wildlife populations are progressively declining, these populations are becoming more and more similar to ex situ populations. In addition to small-population management, metapopulation management also includes (wild-to-wild) translocations and the (re) introduction of captive-bred animals.

In short, animals are constantly on the move from one context to another worldwide. They travel back and forth between ex situ and in situ conservation. Consequently, processes of de-domestication (or re-wilding) alternate with processes of re-domestication (or de-wilding).[2] Animals are becoming increasingly dependent on transitional environments that are neither wild nor domestic. Mixed category environments are becoming more and more prevalent. However, in situations of transition from wildness to domesticity and vice versa, both animal ethics and environmental ethics must also undergo transition.

In animal ethics, we need to develop an alternative to the utilitarian (animal welfare) approach of Peter Singer and the deontological (animal rights) approach of Tom Regan. Within these traditional approaches, animals are usually considered as discrete entities, apart from their environment or habitat and their relationships with other animals, including human animals.[3] Insofar as their environment is taken into account at all, it is the "wild" environment that is seen as the only truly authentic environment from which the welfare, integrity, and intrinsic value of animals should be measured. This individualistic and zoocentric perspective is too narrow to do justice to situations of transition that are becoming ever more dominant.

Whereas traditional animal ethics is typically blind to the environment or habitat of animals, environmental ethics suffers from an opposite deficiency. Here the focus is not on individual animals but on their habitat, biotope, or ecosystem. Both the (individualistic) animal ethics and the (holistic) environmental ethics lose sight of the interaction between animals and their environment. Because species protection policies in mixed category environments should reflect caring for animal welfare *and* animal habitats, the dichotomy between zoocentrism (protecting the welfare of individual animals) and ecocentrism (preserving nature as self-sustaining ecosystems) should be overcome.

In this chapter, we will put forward virtue ethics as a promising alternative to the traditional animal welfare and animal rights approaches, an alternative that can also help to overcome the dichotomy between animal ethics' zoocentrism and environmental ethics' ecocentrism.[4] We will deal with three theories within virtue ethics that are important for an ethics in transition: the theory of Alasdair MacIntyre on the relationship between virtues and practices; the ethics of care of Carol Gilligan, Joan Tronto, and many others; and the capability approach of Martha Nussbaum.

Virtues and Practices

Around 1980, a revival of ancient virtue ethics unfolded that can be seen as a response to the growing disillusionment in the West with the moral and political legacy of the Enlightenment.[5] Enlightenment thinkers like Immanuel Kant and Jeremy Bentham were charged with having replaced the old and rich tradition of the virtues with a new and severely impoverished moral vocabulary.

A case in point is Alasdair MacIntyre. In his 1981 book *After Virtue: A Study in Moral Theory*, MacIntyre argued that an authentic moral life could neither be based on the seemingly exact calculation of costs and benefits (against utilitarianism), nor on the proper application of principles and rules to dilemmatic situations (against deontology). Moral life is not a matter of calculation or rule following, he insists, but of the exercise of the virtues.

MacIntyre attempts to clarify the core concept of virtues by linking them to "practices" or well-marked domains of cooperative activity.

When classical authors like Homer and Aristotle talk about virtues, they often refer to the qualities that are required to participate and excel in such "practices" as warfare, gymnastic games, flute playing, poetry, and geometry. MacIntyre presents a rather technical definition:

> By a "practice" I am going to mean any coherent and complex form of socially established cooperative human activity through which goods internal to that form of activity are realized in the course of trying to achieve those standards of excellence which are appropriate to, and partially definitive of, that form of activity, with the result that human powers to achieve excellence, and human conceptions of the ends and goods involved, are systematically extended. (1981, 187)

As other examples of practices, MacIntyre mentions football, chess, architecture, farming, scientific research, historiography, politics, the management of households, and portrait painting. He also gives a first, tentative definition of a virtue, stating, "A virtue is an acquired human quality the possession and exercise of which tends to enable us to achieve those goods which are internal to practices and the lack of which effectively prevents us from achieving any such goods" (1981, 191).

MacIntyre distinguishes the "internal goods" of a practice from its "external goods" (prestige, wealth and power), which are only contingently connected to it. In others words: football is ideally about achieving excellence in the game of football (which has to be shown in the competition between football clubs), not about realizing high salaries for the players or boosting the quotation of the club's stock at the stock exchange. While most virtues are defined for specific practices, MacIntyre also holds that there are a few virtues that are vital for achieving the "internal goods" of almost *any* practice, which he indentifies as the *moral* virtues, in particular courage, honesty, and justice.

A Typology of Animal-Human Practices

MacIntyre's theory enables us to solve the problem, typical for both the utilitarian and the deontological approach within animal ethics, that animals are considered as discrete entities, apart from their environment and their relationships with other animals, including human animals. Inspired by MacIntyre's theory, Michiel Korthals (2002) has drawn up a typology of various human-animal practices, each with their own internal goods and associated virtues. Korthals made use of Colin Spedding's book *Animal Welfare* from 2000, in which seven categories of animals whose welfare should be our concern are delineated.

- *Farm livestock* such as cattle, goats, sheep, pigs, and poultry.
- *Companion animals (including pets)* such as dogs, cats, hamsters, and rabbits.
- *Captive animals* from zoos and menageries such as large cats, elephants, monkeys, and bears.
- *Working animals* With respect to this category, there are significant differences between developing and developed countries. In developing countries, very large numbers of animals are still used for transport and traction, cultivation and clearing of land, herding and guarding, and so on. In developed countries, machines have displaced most of these animals. Here animals are increasingly used in therapeutic settings. An example is equitherapy, therapeutic horseback riding for individuals with various diagnoses such as autism, multiple sclerosis, and cerebral palsy.
- *Animals used for sport, recreation, and entertainment* This is a very diverse category and includes practices ranging from racing, jumping, riding, and showing to hunting, shooting, fishing, and fighting. Performing animals from circuses and dolphinaria also fall within this category.
- *Wild animals affected by humans* Apart from the killing of wild animals for sport or pest control, Spedding also mentions the need for population control (by culling, relocation, etc.) in the interests of conservation of animals or habitats. He further mentions schemes to rescue toads and frogs and to prevent them from being crushed on busy roads.
- *Animals used in experiments* such as rats, mice, monkeys, dogs, and cats.

To develop a typology of human-animal practices, we will use Spedding's categorization with some amendments (see Korthals 2002, 134).[6] We will subsume working animals under farm livestock, even though this means that some animals such as guide dogs for the blind will be excluded. Furthermore, we will confine the diverse category of "Animals used for sport, recreation, and entertainment" to performing animals, especially animals kept in circuses, a restriction which has to do with the occasion for conducting this study (see table 6.1).[7]

In this diagram we have tentatively ordered the various human-animal practices on a continuum of increasing human dominance over animals—with wildlife parks in which human influence is reduced as much as possible at one extreme and laboratories for animal experiments where human control is at its maximum at the other.

Table 6.1
Human-Animal Practices—From Wildness to Domesticity

Practice	Wildlife park	Zoo	Circus	Household	Farm	Laboratory
Relevant goods	Conservation Restoration	Education Recreation Science Conservation	Entertainment Training Art	Entertainment Company	Food Fiber	Drugs Therapies Science

Source: Keulartz and Swart 2009

Virtues and Care

In his 1999 book *Dependent Rational Animals*, Alasdair MacIntyre made a fundamental change in his position. In this book he criticizes the mainstream of Western ethics, including his own previous view, for overstating the powers of reason and the pursuit of autonomy and for disregarding the dependent and animal sides of human nature. Whereas previously MacIntyre tried to ground the virtues without any reference to what he called Aristotle's "metaphysical biology," he now claims that ethics without biology is impossible (see Laitinen 1999). This fundamental change goes hand in hand with the introduction of the themes of dependence and vulnerability. With respect to these themes, Thomas Aquinas is more relevant to MacIntyre than Aristotle, who puts too much stress on self-sufficiency and "manly" virtues. Besides virtues that enable us to achieve excellence in performing practices, we also need virtues to sustain those relationships that are necessary for us because of our dependence and vulnerability—virtues such as generosity, empathy, and beneficence.

Such virtues are the main subject of an "ethic of care" that sees moral actors as members of a network of relationships on whose continuation they all depend. According to Carol Gilligan, this ethic is rooted in a specific view of the world as one "of relationships . . . where an awareness of the connection between people gives rise to recognition of a responsibility for one another, a perception of the need for response" (Gilligan 1982, 30). Referring to MacIntyre's *After Virtue*, Joan Tronto has suggested that care is perhaps best thought of as a practice. "To call care a practice implies that it involves both thought and action, that thought and action are interrelated, and that they are directed toward some end" (Tronto 1993, 108). The internal good, to use MacIntyre's term, of practices of care is sustaining relationships. Tronto offers a broad definition of care "as a species activity that includes everything that we do to maintain, continue and repair our 'world' so that we can live in it as well as possible. That world includes our bodies, our selves and our environment all of which we seek to interweave in a complex life-sustaining web" (103). Such a broad vision on care opens a promising perspective on an alternative approach to animal ethics, one in which the main moral responsibilities concern sustaining the vitally important relationships of animals not only with other animals (including human animals) but also with their environment.

Usually, considerations of care are thought to apply more to domestic than to wild animals, because domestic animals are dependent on human beings for their welfare and subsistence, whereas truly wild animals are independent and do not need humans for food, shelter, veterinary assistance, and so on. However, the lack of direct or indirect relationships between humans and wild animals does not imply the absence of ethical obligations toward wild animals, because the act of caring should include caring "for the environment, as well as for others" (Tronto 1993, 103). Caring for wild animals means that we should make efforts to sustain their essential relationships with the environment on which they depend for their survival.

Jac. A. A. Swart (2005) has called this type of care "non-specific care" because it is not directed at the individual wild animal and its specific individual needs, as is the case in caring for particular domestic animals. For the latter, Swart has suggested the term "specific care." Specific care has an affective connotation, especially in the case of domestic animals, whereas nonspecific care implies a more distant attitude. Nonspecific care is usually directed at the level of populations or ecosystems and consists of measures that make it possible for wild animals to live their own lives, as wild. The development of ecological networks in densely populated areas in order to give wild animals the opportunity to migrate naturally is an example of nonspecific care. This type of care does not prevent suffering caused by natural conditions, since such conditions are a fact of life in the wild.

As we have argued in the introduction, animals are increasingly dependent on transitional environments that are neither wild nor domestic. Because mixed category environments are becoming more and more prevalent, specific and nonspecific care must not be seen as mutually exclusive concepts. Animals in processes of de-domestication or re-domestication do not simply cross a distinct dividing line between specific and nonspecific care; they do not walk from a moral domain of individual care to one of concern for the ecological whole (Klaver et al. 2002). Instead of a clear-cut borderline between specific and nonspecific care, there is a broad continuum, an intermediate zone in which it is not a question of "either-or" but of "less or more." Our obligations of care should vary depending on the direction of the transition along this continuum from wildness to domesticity (figure 6.1).

Virtues and Capabilities

An important question that an alternative approach to animal ethics has to answer is how to find the optimal balance between specific and non-

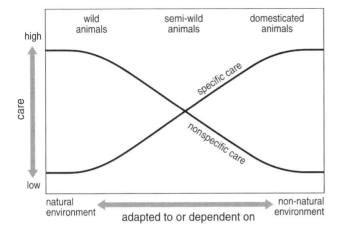

Figure 6.1
Concepts of care along wildness-domesticity continuum (Swart 2005)

specific care for the various human-animal practices along the wildness-domestic continuum. To answer this question Martha Nussbaum's "capability approach" points us in the right direction.[8] By grounding virtue ethics in capabilities, or "substantive freedoms," Nussbaum (2000) has freed virtue ethics of its problematic teleological character. A virtuous person will recognize, protect, and promote the capabilities that help other individuals to flourish, although those capabilities are not pointed toward a definite goal. Nussbaum has proposed a list of ten central human capabilities, which she also uses to map out some basic political principles that can guide law and public policy in dealing with animals.[9]

In her most recent book, *Frontiers of Justice*, Nussbaum (2006) has applied the capability approach to three urgent problems of justice neglected by most current theories (such as John Rawls's theory of justice): justice for the handicapped and disabled, global justice across national boundaries, and justice for nonhuman animals. In the part on nonhuman animals—entitled "Beyond 'Compassion and Humanity'"— Nussbaum sketches the contours of an animal ethics that provides better guidance in dealing with animals than the traditional (utilitarian and deontological) approaches in which animals are usually seen as discrete entities, apart from their environment, a perspective that is too narrow to do justice to the growing number of transitional and mixed category environments.

Nussbaum's capability approach enables us to take the dynamic interaction between animals and their environment much more seriously. According to this approach, the welfare of animals should not be measured against "the one and only true" context of wildness, but must instead be measured against the possibilities an environment offers animals to actually display their basic natural capabilities. If appropriate care is given, animals can flourish in less natural and more human environments. What matters to the welfare of animals is the presence of sufficient opportunities to employ their natural capabilities, not the naturalness of the environment (see Musschenga 2002, 179).

In *Frontiers of Justice*, Nussbaum is looking for a balance between Swart's specific and nonspecific care distinctions. We should preserve the habitat of wild species such as tigers, but we should at the same time take care that they can flourish in captivity. Nussbaum also enables us to make a distinction within the category of specific care between species-specific care and care for the individual animal. Although the focus of the capabilities approach is on the well-being and dignity of the individual creature, Nussbaum rejects "the view that species membership in itself is of no moral relevance and that all moral relevance lies in the capacities of the individual" (Nussbaum 2006, 358). According to the kind of paternalism she champions with respect to domesticated animals and wild animals in captivity, "the touchstone should be a respectful consideration of the species norm of flourishing and a respectful attention to the capacities of the individual" (378).[10] In short, we need to look for the appropriate mix of nonspecific, species-specific, and individual care within the various human-animal practices. In the remainder of this paper we will try to do so for three of these practices: wildlife parks, zoos, and circuses.

Wildlife Parks and Nonspecific Care

Like most animal ethicists, Nussbaum attaches moral weight to the possibility for animals to enjoy sovereignty. She endorses "the idea that species autonomy . . . is part of the good for nonhuman animals" (2006, 375). Part of what it is to flourish, for a creature, "is to settle certain very important matters on its own, without human intervention, even of a benevolent sort" (373). This last qualification is important once we realize that life in the "state of nature" is often "solitary, poor, nasty, brutish, and short," to use Hobbes's famous words. Compared to animals in some human-animal practices, morbidity and mortality rates among animals in the wild are rather high. But this should be no reason to take the animals into preventive custody, as it were, although there are some ethicists (mostly from the utilitarian camp) who favor such a policy. To Nussbaum "the very idea of a benevolent despotism of humans over animals, supplying their needs, is morally repugnant" (373). Her criterion for animal flourishing is not the degree of welfare and lack of pain or inconvenience but the extent to which essential capabilities can be realized. In principle, animals can pursue their own flourishing best when left to their own devices. But this, Nussbaum adds, is only the beginning, not the end of the story.

Although animals in the "wild" should preferably live sovereign and autonomous lives, unaffected by human interference, one ought to realize that such a state of nature is becoming more and more of an exception. The environments on which animals depend for their survival are being increasingly disturbed or destroyed, and their opportunities for nutrition, shelter, and free movement are in constant decline. In this situation, Nussbaum advocates what we have called "non-specific care"—we should not leave the animals to their fate but we should take measures to preserve or restore their habitat. "Even a person who wanted to deny that we had responsibilities to animals in the 'wild' before this century ought to grant that our pervasive involvement with the conditions of animal flourishing gives us such responsibilities now" (2006, 374). Nussbaum gives the following example: "It is no use saying that we should just let tigers flourish in their own way, given that human activity ubiquitously affects the possibilities for tigers to flourish, and indeed, to live at all. This being the case, the only decent alternative to complete neglect of tiger flourishing is a policy that thinks carefully about the flourishing of tigers and what habitat that requires, and then tries hard to create such habitats" (375).

This type of nonspecific care becomes manifest in the establishment of a global network of protected areas. Currently, the World Database on Protected Areas comprises over 105,000 sites covering an area of 19.7 million kilometers, including all nationally designated (e.g., national parks, nature reserves) and internationally recognized protected areas (e.g., UNESCO World Heritage Sites, Ramsar Wetlands of International Importance).

In protected areas, care is focused on the habitat of animals, not on their individual welfare. This often leads to public consternation and political commotion. A case in point concerns the recent Dutch policy of introducing domesticated and semiwild large herbivores in newly developed nature areas where the herbivores are subject to de-domestication. The management policies of de-domestication, which entail minimizing supplementary feeding and veterinary assistance, have been most controversial. But, the proponents of these policies argue, human intervention on behalf of individual animals will undermine their ability to fend for themselves and lead autonomous lives (ICMO 2005). In short, the effective management of protected areas and the animals therein requires what Nussbaum calls a "delicate balancing" of various factors involved: "As in the foreign-aid case, the best form of aid is a form that preserves and enhances autonomy, rather than increasing dependency. It would be a bad result if all animals ended up in zoos, completely dependent on human arrangements" (Nussbaum 2006, 375).[11]

However, because of ongoing fragmentation of the landscape, nitrogen deposition, climate change, and many other human-induced changes, nonspecific care is no longer sufficient as animals are increasingly on the move and becoming more and more dependent on care in conditions of captivity—hence the new role of zoos in species preservation today.

Zoos and Species-Specific Care

"Captivity for Conservation" turns out to be the basic mission of many modern zoos.[12] Here captive propagation and repatriation programs are being developed to aid the conservation of animals in the wild. A number of species, such as the Asian Wild Horse, the Black-footed Ferret, the Arabian Oryx, the Red Wolf, Pere David Deer, and the California Condor, owe their survival to the cooperative efforts of modern zoos (Hutchins, Dresser, and Wemmer 1995).[13]

The tiger also provides a telling example. According to recent research, captive tigers are indispensible for the conservation of tigers in the wild. The numbers of wild tigers has decreased dramatically, from over one hundred thousand in the 1900s to as few as three thousand. Some subspecies are extinct while others persist only in zoos. With fifteen to twenty thousand individuals, captive tigers nowadays outnumber their wild relatives five to seven times. As a consequence of the larger population size and of breeding strategies, captive tigers have retained genetic diversity unobserved in their wild counterparts (Luo et al. 2008).

Under such circumstances Nussbaum argues for a double strategy. We should not only try to preserve or restore at least some part of the original habitat of endangered species, but we should also make "intelligent and careful" use of zoos. "Many animals will do better in an imaginative and well-maintained zoo than in the wild, at least in present conditions of threat and scarcity" (Nussbaum 2006, 376).[14]

In zoos, the focus shifts from (nonspecific) care for the environment to specific care, especially species-specific care. Animal flourishing depends on the options and opportunities to perform species-specific activities. There are, however, limitations to the extent to which these activities can be carried out in zoos. It may be true that there is a strong tendency toward the "naturalization" of zoos, but this process runs up against limits. Zoos cannot include the reproduction of natural contingencies. Some forms of predatory behavior, such as chasing and killing prey, cannot realistically be simulated in captivity. Likewise, in the absence of predators, some forms of prey behavior, such as vigilance, may not be exhibited at appropriate levels in captivity.

In short, captivity usually deprives wild animals of the necessity and opportunity to pursue the tasks of survival, such as finding food and avoiding animals. Heini Hediger, one of the founding fathers of the science of "environmental enrichment" considered this lack of occupation of the captive animal as one of the most urgent problems of zoological gardens. Quite a lot of studies show that animals prefer to work for their food, rather than to be fed *ad libitum*. Especially mammals are unsuited to an existence in which no effort on their part is required to meet their basic needs (Kreger, Hutchins, and Fascione 1998).

A solution to this problem is "substitution." Animals, particularly mammals, are flexible enough to modify their behavior to suit a wide range of situations and to substitute one form of action for another depending on the facilities available. Trevor Poole (1998) refers to studies that show that the absence of a forest full of interesting foods may be of little concern to a chimpanzee or mandrill when the animal has the opportunity to play a computer game. Computer games seem to provide a good alternative for the explorative behavior chimpanzees show when they are searching for food.

In *Frontiers of Justice*, Nussbaum mentions another example of substitution. Modern zoos have to face the problem of allowing the capabilities of predatory animals like tigers to be exercised without actually harming or killing prey animals. The Bronx Zoo has found an answer to this question—instead of giving a tiger a tender gazelle to crunch on, it gives the tiger a large ball on a rope, whose resistance and weight symbolize the gazelle. "Wherever predatory animals are living under direct support and control," Nussbaum concludes, "these solutions seem the most ethically sound" (Nussbaum 2006, 370).

A special case of substitution is what Hediger has called "occupational therapy." "The captive animal," Hediger suggested, "must be given an adequate substitute for the chief occupation of freedom. . . . This substitute can take the form of biologically suitable training and assumes the importance of occupational therapy" (cited in Laule and Desmond 1998, 305). Training offers animals a chance to work for their food, by performing certain tasks or behaviors for a food reward.

Training is also relevant for the relationship between the animal and its caregiver, which is of critical importance for many species in captivity. The well-being of zoo animals is greatly enhanced if there is a trusting, amiable relationship with their keepers. Studies have shown that the quality of this relationship is reflected in reproductive success in captivity. The trust between animals and keepers can be reinforced by training animals to follow commands and teaching them simple routines using positive reinforcement. As Trevor Poole has noticed, zoos are reluctant to use training, because they fear to be accused of not being serious or of turning into circuses, "but training undoubtedly improves the welfare of the animal and also makes routine examination and veterinary treatment much easier and less stressful" (Poole 1998, 91).[15]

Circuses and Individual Care

When we turn to circus animals, the balance of care will shift once more, from species-specific care to individual care. Nussbaum argues that both types of care are important for animals to flourish. Her perspective is individualistic in making the living creature—not the group or species—the basic subject of justice, but at the same time she rejects the view that all moral relevance lies in the capacities of the individual. The capabilities approach should give a species-specific account of basic capabilities without losing sight of the capacities and personality of the individual animal. A respectful consideration of the species norm of flourishing should go hand in hand with a respectful attention to the capacities of the individual.

In animal ethics, the capacities and character of individual animals generally have attracted far less attention than the species-specific norm of flourishing. To gain insight in the prospects and problems of individual care for performing animals, including circus animals, it is useful to have a look at the work of two philosophers who have been theoretically and practically engaged in animal training: Donna Haraway and Vicky Hearne.

Donna Haraway, a path-breaking feminist philosopher, best known for her 1991 essay "The Cyborg Manifesto," has recently been focusing her attention on human-animal relationships. In 2003, she published *The Companion Species Manifesto*, and in 2008 her book *When Species Meet* came out. What holds this abundant and complex book together— "besides the author's over-the-top intellectual power" (Peterson 2008, 611)—is Haraway's relationship with her Australian shepherd Cayenne, with which she does agility work.

Haraway is less interested in Jeremy Bentham's famous question "Can animals suffer?" than in the questions: Can animals play? Or work? To better understand and nurture our responsibilities for animals Haraway feels that Karl Marx's category of labor is more helpful than the Enlightenment's category of rights. If, however, we are going to include animals within the regime of lively capital ("biocapital"), Haraway contends, we should reconsider Marx's theory of value, adding another notion to Marx's notions of use-value and exchange-value: "encounter-value."

Haraway illustrates the importance of labor or work in our relationships with animals by summing up some of the tasks done by herding dogs, livestock-guarding dogs, attack dogs, laboratory dogs, airport security dogs, drug and bomb sniffers, and so on. Today, dogs are increasingly involved in therapeutic work, including "warning of epileptic seizures, detecting cancer, guiding the blind, serving as aides for the hearing impaired and the wheelchair-bound and as psychotherapeutic aides for traumatized children and adults, visiting the aged, aiding in rescues in extreme environments, and more" (Haraway 2008, 57).

Haraway is of course aware that many animal ethicists and animal rights activists find this kind of work morally repugnant, often because they consider domestication as original sin and feel that only animals in the wild can live up to their *telos*. But training animals does not imply submission and oppression; quite the contrary, it requires two-way interaction. Training relies on teamwork; trainer and trainee have to listen to one another if an act or exercise is going to succeed. Here, Haraway refers to the work of the Belgian philosopher and psychologist Vinciane Despret, who has reframed domestication using the notion of "anthropo-zoo-genetic practices," in which humans and animals co-shape their historically situated relationships. Despret points to the notion of "iso-praxis" used by the French ethologist Jean-Claude Barrey to describe the interaction between horse and rider. According to Barrey, talented riders behave and move like horses: "They have learned to act in a horse-like fashion, which may explain how horses may be so well attuned to their humans, and how mere thought from one may simultaneously induce the other to move. Human bodies have been transformed by and into a horse's body" (Despret 2004, 115). According to Despret, the questions "Who influences and who is influenced?" can receive no clear answer. "Both, human and horse, are cause and effect of each other's movements. Both induce and are induced, affect and are affected. Both embody each other's mind" (2004, 115).

Haraway is also influenced by the work of Vicky Hearne. In a section of *The Companion Species Manifesto* entirely devoted to this animal trainer, philosopher, and poet, who died in 2001, Haraway claims that Hearne "is in love with the beauty of the ontological choreography when dogs and humans converse with skill, face-to-face" (Haraway 2003, 51). Hearne, who also put much emphasis on mutuality and reciprocity in the working relationships between humans and animals, fiercely opposed the animal rights discourse. Hearne (1992) feared that animal rights activists, by branding training of any sort as torture, could deprive animals of the satisfaction they experience from work. She disagrees with the definition of "happiness" as synonym for "pleasure" and as antonym for "suffering." To better understand "animal happiness" we should trade in Bentham for Aristotle, who considered happiness as activity in accordance with the highest excellence. Happiness is about a sense of personal achievement, like the satisfaction felt by a good wood-carver or a dancer or a poet or an accomplished dressage horse. "This happiness, like the artist's, must come from something within the animal, something trainers call talent, and so cannot be imposed on the animal" (Hearne 2007, 204).

From the work of Haraway and Hearne, it can be concluded that responsibility for animals in working relations should be considered—to use Haraway's phrase—as "response-ability," the ability to listen to animals and to meet their needs. Whether a specific setting is morally acceptable or culpable, so Despret claims, depends on the animal's pos-sibilities of "resistance" when its wishes fall on deaf ears and it is forced to compliance and docility.[16]

On the one hand, it is obvious that the possibilities for nonspecific care in a circus setting are almost absent, while the possibilities for species-specific care are fairly limited. The possibilities for individual care and attention, on the other hand, seem relatively large. One should of course, on a case-by-case basis, try to assess whether the happiness that results from the labor of training can compensate for the lack of non-specific care and the deficit of species-specific care.

Conclusion

Nowadays, animals are constantly on the move; they are increasingly located in rapidly changing contexts that escape simple categorizations in terms of "wild" and "domesticated." Humans and animals are embedded in a broad diversity of relationships, which includes processes of de-domestication or re-domestication. Likewise, the environments in which animals are found are themselves hybrid and constantly changing.

Standard ethical views about animals are ill equipped to deal with these fluid situations; both (individualistic) animal ethics and (holistic) environmental ethics lose sight of the interaction between animals and their environment. Because of its context sensitivity, virtue ethics is a promising alternative to these standard ethical approaches. Nussbaum's capability approach in particular allows for a better understanding of the ways our moral responsibilities toward animals will vary according to the different relationships they have with us. This approach enables us to measure the moral adequacy of various human-animal practices against the options and opportunities they offer animals to exhibit their natural and individual capabilities.

Depending on its position along the wildness-domesticity axis, a specific human-animal practice should show a balanced mix of nonspecific, species-specific, and individual care. As we have shown with respect to three practices, the balance between these types of care will gradually shift if we move from wildness to domesticity: in protected nature areas, the weight is on nonspecific care for the habitat of animals. In zoos, the weight is on species-specific care, while performing animals will predominantly depend on individual care.

We hope that the idea of a broad continuum of human-animal practices together with the differentiation between types of care and the emphasis on animal capabilities will enable policymakers and practitioners to give a new meaning to the concept of interspecies justice in the transitional, mixed category environments in which animals increasingly will find themselves in an age of global climate change.

Acknowledgments

The authors thank Clare Palmer for her useful comments on an earlier draft of this chapter.

Notes

1. In 2008, the Stratigraphy Commission of the Geological Society of London decided, by a large majority, to examine the possibility of formally including this term in the Geological Time Scale (Zalasiewicz et al. 2010).

2. This can be illustrated by the wild horse protection policies in the United States and the Netherlands (Reed 2008). In both countries policies are aimed at restoring self-sustaining ecosystems. But whereas Dutch conservation policy favors releasing domestic breeds into Dutch nature reserves as ecological surrogates for extinct wild horses, wild horse protection policies in the United States favor removing them from public rangelands, because they are not considered an indigenous species. In Dutch nature reserves animals are in transition from domesticity to wildness (de-domestication), while wild horses in the United States have to make a transition from wildness to domesticity—they are being removed from public rangelands and kept in holding facilities awaiting adoption (re-domestication).

3. For an early attempt to move away from utilitarian and deontological approaches to animal ethics and to focus instead on contexts and relationships, see Palmer (1995).

4. We are aware that we don't do full justice to specific authors or positions. For instance, we believe that Singer's type of preference utilitarianism comes rather close to Nussbaum's capabilities approach. Singer once responded to Martha Nussbaum in the following way: "The alliance between utilitarianism and the capabilities approach is, I believe, not so much like that between, say, Britain, the United States and the Soviet Union in the second World War, but rather more like that between the United States and Australia today." See "A Response to Martha Nussbaum" (http://www.utilitarian.net/singer/by/20021113 .htm). Although we have to admit that our account of animal and environmental ethics is somewhat indiscriminate, we do think that it does capture the broad, general trends: the focus of mainstream animal ethics is on individual animals, whereas in environmental ethics animals are mainly considered as parts of ecological wholes. This picture certainly represents a simplification, but not an oversimplification.

5. In fact, this revival can be traced back further, typically to Anscombe (1958).

6. In this typology, at least two categories are absent, scavenging and feral animals (see Palmer 1995). Scavenging animals live alongside humans and do not participate in animal-human practices. The case of feral animals seems to be more complex (see also Palmer 2003). These animals are sometimes taken care of in animal shelters or holding facilities awaiting adoption.

7. This chapter draws substantially on a paper that was written on occasion of the fierce discussions about the condition of wild animals in circuses that are taking place in the Netherlands in recent years. Pressured by parliament, the minister of Agriculture, Nature and Food Quality commissioned a large-scale investigation of the welfare of wild circus animals. The minister also met the wish of parliament to include the issue of the intrinsic value of wild circus animals. We were asked to prepare a report on this issue (Keulartz and Swart 2009).

8. The core characteristic of the capability approach is its focus on what people are effectively able to do and to be. It considers these capabilities as constitutive parts of development, and poverty as capability deprivation. See Holland, chapter 7, this volume, on the basic conception of the capability approach and the core capability set.

9. See Schlosberg, chapter 8, this volume.

10. Schlosberg's criticism (chapter 8, this volume) of Nussbaum's "individualist focus" is somewhat simplistic and needs to be qualified.

11. Nussbaum has contradicting threads running through her account. Later in *Frontiers of Justice* she seems to end up with a view that does entail considerable intervention in nature on behalf of sick animals and animals that are victims of predation. Because nature is never going to provide all species with "cooperative and mutually supportive relations," Nussbaum suggests, there should be a "gradual supplanting of the natural by the just" (Nussbaum 2006, 400).

12. A major milestone in this development was the first World Zoo Conservation Strategy, published in 1993 in the wake of the Rio Earth Summit and the inception of the Convention on Biological Diversity, in which zoos and aquaria were recognized as the main institutions for ex situ conservation of wild animal species.

13. A problem with trying to introduce zoo animals to the wild is that the behavior they have acquired in captivity is largely irrelevant for their survival in the natural environment. Since the adoption of the Convention on International Trade in Endangered Species (CITES) in 1974, most zoo animals, like most circus animals, are not wild-caught but captive-born and captive-bred.

14. In "Animal Thinking and Animal Rights," Nussbaum (forthcoming) discusses an interesting example of a balancing between specific and nonspecific care. Here Nussbaum argues that one should treat African and Asian elephants differently. Species protection policies with respect to African elephants should focus on preserving or restoring their habitat, while these policies are not adequate for Asian elephants because they are far more endangered than African elephants, and also because they have been working in symbiosis with humans to a greater extent than have African elephants. So, for Asian elephants Nussbaum argues for an "intelligent, species-sensitive paternalism" within zoos.

15. Modern circuses not only could be important to zoos as suppliers of training programs, they also could play a modest role in dealing with "the perplexing surplus-animal problem" (Norton 1995, xxiii). Reallocation to circuses could be a partial alternative for culling of surplus animals that are an unavoidable byproduct of captive breeding programs for future reintroduction of such animals into the wild.

16. In this respect the Code of Conduct for Animals by the European Circus Association (ESA 2007) is clear: "All animal training must be based on operant conditioning and the use of positive reinforcement and repetition of desired behaviors. Training should showcase individual animals' natural behaviors and athletics."

References

Anscombe, G. E. M. 1958. Modern Moral Philosophy. *Philosophy* 33 (124): 1–19.

Despret, V. 2004. The Body We Care for: Figures of Anthropo-zoo-genesis. *Body & Society* 20 (2–3): 111–134.

European Circus Association (ESA). 2007. Code of conduct for animals. http://www.europeancircus.info/ECA/index.php?option=com_content&task=view&id=83&Itemid=63 (accessed May 15, 2009).

Gilligan, C. 1982. *In a Different Voice. Psychological Theory and Woman's Development*. Cambridge, MA: Harvard University Press.

Haraway, D. 2003. *The Companion Species Manifesto: Dogs, People, and Significant Otherness*. Chicago: Prickly Paradigm Press.

Haraway, D. 2006. Encounters with Companion Species: Entangling Dogs, Baboons, Philosophers, and Biologists. *Configurations* 14: 97–114.

Haraway, D. 2008. *When Species Meet*. Minneapolis: University of Minnesota Press.

Harris, J. A., R. J. Hobbes, E. Higgs, and J. Aronson. 2006. Ecological Restoration and Global Climate Change. *Restoration Ecology* 14: 170–176.

Hearne, V. 1987. *Adam's Task. Calling Animals by Name*. London: Heinemann.

Hearne, V. 1992. What's Wrong With Animal Rights: Of Hounds, Horses and Jeffersonian Happiness. In *The Best American Essays*, ed. Susan Sontag, 199–208. New York: Houghton Mifflin.

Hearne, V. 2007. *Animal Happiness. A Moving Exploration of Animals and Their Emotions*. (With a new introduction by Elizabeth Marshall Thomas.) New York: Skyhorse Publishing.

Hutchins, M., B. Dresser, and C. Wemmer. 1995. Ethical Considerations in Zoo and Aquarium Research. In *Ethics on the Ark: Zoos, Animal Welfare and Wildlife Conservation*, ed. Bryan Norton, Michael Hutchins, Elizabeth Stevens, and Terry Maple, 253–276. Washington, DC: Smithsonian Institution Press.

ICMO. 2005. *Reconciling Nature and Human Interest. Advice of the International Committee on the Management of Large Herbivores in the Oostvaardersplassen*. The Hague/Wageningen, Netherlands: WING Proces Consultancy.

Keulartz J., and J. A. A. Swart. 2009. "De intrinsieke waarde van dieren in performancepraktijken [The intrinsic value of performing animals]." Report 216, Animal Sciences Group, Wageningen University and Research Centre.

Klaver, I., J. Keulartz, H. van den Belt, and B. Gremmen. 2002. Born to Be Wild. *Environmental Ethics* 24 (1): 3–21.

Korthals, M. 2002. A Multi-Practive Ethics of Domesticated and "Wild" Animals. In *Pragmatist Ethics for a Technological Culture*, ed. J. Keulartz, M. Korthals, M. Schermer, T. Swierstra, 127–141. Dordrecht: Kluwer Academic Publishers.

Kreger, M. D., M. Hutchins, and N. Fascione. 1998. Context, Ethics, and Environmental Enrichement in Zoos and Aquariums. In *Second Nature: Environmental Enrichment for Captive Animals*, ed. D. Shepherdson, J. Mellen, and M. Hutchins, 59–82. Washington, DC: Smithsonian Institution Press.

Laitinen, A. 1999. A Review of Alasdair MacIntyre, Dependent Rational Animals: Why Human Beings Need the Virtues. *Associations, Journal for Legal & Social Theory* 5 (1): 142–150.

Laule, G., and T. Desmond. 1998. Positive Reinforcement Training as an Enrichment Strategy. In *Second Nature: Environmental Enrichment for Captive Animals*, ed. D. Shepherdson, J. Mellen, and M. Hutchins, 302–313. Washington, DC: Smithsonian Institution Press.

Luo, S.-J., et al. 2008. Subspecies Genetic Assignments of Worldwide Captive Tigers Increase Conservation Value of Captive Populations. *Current Biology* 18: 592–596.

MacIntyre, A. 1981. *After Virtue: A Study in Moral Theory*. Notre Dame, IN: Notre Dame University Press.

MacIntyre, A. 1999. *Dependent Rational Animals: Why Human Beings Need the Virtues*. Chicago: Open Court.

Musschenga, A. W. 2002. Naturalness: Beyond Animal Welfare. *Journal of Agricultural & Environmental Ethics* 15: 171–186.

Norton, B. G. 1995. Preface. In *Ethics on the Ark: Zoos, Animal Welfare, and Wildlife Conservation*, ed. B. G. Norton, M. Hutchins, E. F. Stevens, and T. L. Mapple, xxi–xxii. Washington, DC: Smithsonian Institution Press.

Nussbaum, M. 2000. *Women and Human Development: The Capabilities Approach*. Cambridge, UK: Cambridge University Press.

Nussbaum, M. C. 2006. *Frontiers of Justice: Disability, Nationality, Species Membership*. Cambridge, MA: Belknap Press of Harvard University Press.

Nussbaum, M. C. Forthcoming. Animal Thinking and Animal Rights. In *The Oxford Handbook of Animal Ethics*, ed. Tom Beauchamp and R. G. Freay. Oxford: Oxford University Press.

Palmer, C. 1995. *Animal Liberation, Environmental Ethics and Domestication*. Oxford Centre for the Environment, Ethics & Society (OCEES) Research Paper No 1. Oxford, UK.

Palmer, C. 2003. Placing Animals in Urban Environmental Ethics. *Journal of Social Philosophy* 34 (1): 64–78.

Peterson, A. 2008. Donna J. Haraway, When Species Meet. *Journal of Agricultural & Environmental Ethics* 21: 609–611.

Poole, T. B. 1998. Meeting a Mammal's Psychological Needs. In *Second Nature*, ed. D. J. Shepherdson, J. D. Mellen, and M. Hutchins, 83–97. Washington, DC: Smithsonian Institute.

Reed, C. M. 2008. Wild Horse Protection Policies: Environmental and Animal Ethics in Transition. *International Journal of Public Administration* 31 (3): 277–286.

Spedding, C. 2000. *Animal Welfare*. Sterling, VA: Earthscan.

Swart, J. A. A. 2005. Care for the Wild. Dealing with a Pluralistic Practice. *Environmental Values* 14 (2): 251–263.

Swart, J. A. A. 2009. Towards an Epistomoloy of Place. In *New Visions of Nature: Complexity and Authenticity*, ed. M. Drenthen, J. Keulartz, and J. Proctor, 197–203. Dordrecht: Springer Verlag.

Tronto, J. C. 1993. *Moral Boundaries. A Political Argument for an Ethics of Care*. New York: Routledge.

Zalasiewicz, J., M. Williams, W. Steffen, and P. Crutzen. 2010. The New World of the Anthropocene. *Environmental Science & Technology* 44 (7): 2228–2231.

7

Environment as Meta-capability: Why a Dignified Human Life Requires a Stable Climate System

Breena Holland

What are the prospects for human flourishing on a planet facing dramatic ecological changes brought on by a warming global climate? Scientific research warns of a future in which significant portions of the human population will face more drought, flooding, famine, and disease. These problems are closely related to earth's ecology. As global temperatures rise, changes in weather patterns and severity will disrupt the flow of ecological services that people in many areas of the world now take for granted, such as groundwater recharge and availability of stable land substrates that provide a solid foundation for human homes (Millennium Assessment Panel 2003; Wackernagel et al. 2002; UNEP 2007). When facing the loss of such basic conditions of human living, it might seem naively utopian to even raise the question of what a flourishing life would require. After all, Western society has spent many decades promoting an economy that seeks the much narrower goal of merely satisfying consumer preferences.[1] Now that climate change threatens to put this goal out of reach for millions of people who still lack basic consumer goods, is it not disingenuous to raise the prospect of sustaining a broader and more complex set of human activities and achievements, such as those that might make up a virtuous human life?

Many environmental thinkers have denied that such a life is unattainable. They have argued, for example, that redirecting economic activities can reduce environmental damage. By regionalizing the economy, politics, and culture, it is possible to create a "mutually sustainable future for humans and other life forms . . . in which people live as rooted, active, participating members of a reasonably scaled, naturally bounded, ecologically defined territory, or *life place*" (Thayer 2003, 6). From this perspective, to promote a wider and deeper view of human living is to promote human practices that will demand fewer resources and produce less waste; human flourishing and environmental protection can go hand in hand.

Unfortunately, given what we now know about large-scale biogeo-chemical processes (e.g., the carbon cycle) that connect the personal, social, and economic practices of people in distant places, the prospects for human flourishing in many areas of the world seem increasingly unlikely. On a planet altered by climate change, many important envi-ronmental resources will become even scarcer. Using the environment in a way that enables the most minimal conditions of human living in one place may degrade the environment needed to enable the most minimal conditions of human living in other places. Consider, for example, that in order to sustain a world population of 6.2 billion people—a population count the earth already exceeds—a sustainable level of consumption would permit a per capita income of only $5,100/year (Worldwatch Institute 2010, 6). Put differently, if the entire world were to live on a per capita income of $35,690/year, the planet could only sustain 2.1 billion people, which would require reducing the global population to less than a third of what it is today. What many people in the developed world understand as basic to a good or flourishing human life is only sustainable if much of the underdeveloped world remains impoverished.[2]

The likelihood that climate change will further inequities lends impor-tance to questions about how contemporary society can adapt to a changing global climate while sustaining basic biotic conditions neces-sary for ensuring each individual person opportunities to live a good life. Furthermore, given a historic failure of Western society to forestall the environmental crisis that no society can now escape, it is unclear whether contemporary political and philosophical thinking about government in liberal democratic societies is adequate for understanding and reasoning about the implications of human impacts on global ecological interac-tions.[3] As Dale Jamieson (chapter 9, this volume) claims, the value system contemporary society has inherited from Western philosophy is "inade-quate and inappropriate for guiding our thinking about global environ-mental problems." Much besides economic growth hangs in the balance, for in determining what forms of nonhuman life will survive (and in what condition), these problems will also determine who among humans and nonhumans will benefit from the opportunities for flourishing that functioning ecosystems make possible.[4]

While adapting to climate change may require "that there be people of integrity and character who act on the basis of principles and ideals,"[5] it also requires institutions that create a space for such principles and ideals to matter. This chapter explores Martha Nussbaum's "capabilities approach" to justice as a set of ideas and concepts for thinking and reasoning about how institutions in a politically liberal democratic society should respond to climate change. Nussbaum adopts a neo-Aristotelian conception of what a minimally decent and dignified human life requires, and she articulates this in terms of ten central human capabilities that a just society should protect at a threshold level for each individual person. These capabilities refer to conditions or states of enablement that make it possible for people to do things. They include being able to attain basic material conditions, such as shelter and nutrition, as well as more sophisticated capacities, such as emotional attachments to people and critical reflection about the direction of one's life. By embedding this multidimensional view of what makes people capable of living a life "worthy of the dignity of a human being" within a political theory about what social justice requires, Nussbaum offers a basic theoretical framework for reasoning about the institutional components of a just response to ecological systems degraded by climate change.

In practice, Nussbaum herself does not theorize the multiple ways in which the natural environment and a stable climate system can support or undermine the capabilities she defines as necessary for living a dignified human life. However because a stable climate system provides the necessary ecological context for engaging in personal, social, material, and political relationships that Nussbaum defines as basic to any minimally decent notion of human flourishing, I argue that such a system should be understood as part of a broader "environmental meta-capability" that enables all the capabilities worthy of protection as constitutional entitlements. Given this characterization of justice, I then consider how it can inform the goals of ecosystem restoration. Overall, by incorporating ecological thinking into a view of justice that seeks to enable the flourishing of humans, this chapter aims to provide criteria for evaluating whether social institutions are acting responsibly in an ecologically interconnected world—a world in which institutions carve out a space for virtuous citizens to "think like a planet."[6]

Nussbaum's Capabilities Approach to Justice

In a series of books and articles, Martha Nussbaum has developed a liberal theory of justice—the capabilities approach—that specifically aims to address the special circumstances faced by women, nonhuman animals, disabled citizens, and noncitizens.[7] Central to the vision of a just liberal democratic society that Nussbaum advances is the idea that justice should be defined in terms of people's capabilities to do and achieve different things.[8] By "capabilities," Nussbaum means the conditions or states of enablement that make it possible for people to engage in activities like affiliating with members of their community and to achieve conditions of well-being like being adequately nourished. Capabilities, in other words, are people's real opportunities to pursue goals they value, such as participating in the political decisions that govern one's individual and collective life and engaging in activities that physical health and mobility make possible. Nussbaum argues that without protecting the conditions that enable people to do these things, an individual's right to them as a matter of justice will remain merely formal for at least some members of society. Those who lack these capabilities will lack political protection for the bare minimum of what is necessary for living even a minimally decent human life (Nussbaum 2000, 72–73; 2006, 70–78).

Nussbaum grounds this argument for protecting people's capabilities in a tradition of political thought that emerged from Aristotle, continued on through Karl Marx and J. S. Mill, and can be seen today in her view of what respect for each person's basic human worth requires. Recalling Marx's description of deprivations suffered by a starving man who just "grabs at the food in order to survive," Nussbaum characterizes a life worthy of human dignity as one in which the many social and rational ingredients of human living can make their appearance, such that a person "shapes his or her own life in cooperation and reciprocity with others, rather than being passively shaped or pushed around by the world in the manner of a 'flock' or 'herd' animal" (Nussbaum 2000, 72; for the relevant text in Marx's writings, see Tucker 1978, 89). To live a life that is dignified in this sense, one must be capable of making choices free from the kind of desperate conditions in which one's unique faculties to taste, enjoy, and reason about one's food become irrelevant. For Nussbaum, this requires protecting and providing a minimum (threshold) level of the various conditions and states enablement—the capabilities—that actually put the goals and achievements a person values within the person's reach (Nussbaum 2000, 74).

To determine what specific capabilities warrant protection as a matter of justice, Nussbaum draws on her own effort over a number of years to achieve cross-cultural dialogue and consensus among people in many different countries about what the components of a good human life would involve. While the list of capabilities emerging from this initial effort is subject to revision, as might be required by democratic outcomes that result from sustained and reflective moral argument, the list is drafted as follows (Nussbaum 2000, 78–80):

1. *Life* Being able to live to the end of a human life of normal length; not dying prematurely, or before one's life is so reduced as to be not worth living.

2. *Bodily Health* Being able to have good health, including reproductive health; to be adequately nourished, to have adequate shelter.

3. *Bodily Integrity* Being able to move freely from place to place; having one's bodily boundaries treated as sovereign, i.e., being able to be secure against assault, including sexual assault, child sexual abuse, and domestic violence; having opportunities for sexual satisfaction and for choice in matters of reproduction.

4. *Senses, Imagination, and Thought* Being able to use the senses, to imagine, think, and reason—and to do these things in a "truly human" way, a way informed and cultivated by an adequate education, including, but by no means limited to, literacy and basic mathematical and scientific training. Being able to use imagination and thought in connection with experiencing and producing self-expressive works and events of one's own choice—religious, literary, musical, and so forth. Being able to use one's mind in ways protected by guarantees of freedom of expression with respect to both political and artistic speech, and freedom of religious exercise. Being able to search for the ultimate meaning of life in one's own way. Being able to have pleasurable experiences, and to avoid non-necessary pain.

5. *Emotions* Being able to have attachments to things and people outside ourselves; to love those who love and care for us, to grieve at their absence; in general, to love, to grieve, to experience longing, gratitude, and justified anger. Not having one's emotional development blighted by overwhelming fear and anxiety, or by traumatic events of abuse or neglect. (Supporting this capability means supporting forms of human association that can be shown to be crucial in their development.)

6. *Practical Reason* Being able to form a conception of the good and to engage in critical reflection about the planning of one's life. (This entails protection for the liberty of conscience.)

7. *Affiliation* A. Being able to live with and toward others, to recognize and show concern for other human beings, to engage in various forms of social interaction; to be able to imagine the situation of another and to have compassion for that situation; to have the capability for both justice and friendship. (Protecting this capability means protecting insti-

tutions that constitute and nourish such forms of affiliation, and also protecting the freedom of assembly and political speech.) B. Having the social bases of self-respect and non-humiliation; being able to be treated as a dignified being whose worth is equal to that of others. This entails at a minimum, protections against discrimination on the basis of race, sex, sexual orientation, religion, caste, ethnicity, or national origin. In work, being able to work as a human being, exercising practical reason and entering into meaningful relationships of mutual recognition with other workers.

8. *Other Species*   Being able to live with concern for and in relation to animals, plants, and the world of nature.

9. *Play*   Being able to laugh, to play, to enjoy recreational activities.

10. *Control over One's Environment*   A. *Political*   Being able to participate effectively in political choices that govern one's life; having the right of political participation, protections of free speech and association. B. *Material*   Being able to hold property (both land and moveable goods), not just formally but in terms of real opportunity; and having property rights on an equal basis with others; having the right to seek employment on an equal basis with others; having the freedom from unwarranted search and seizure.

For Nussbaum, justice requires that a threshold level of each of these capabilities be provided and protected for each person. Specifically, she argues that citizens in liberal democratic societies should think of these capabilities as constitutional guarantees they have a right to demand from their governments.

Characterizing justice in terms of what people are actually able to do and achieve is especially helpful for understanding how ecological conditions and climate change bear on the possibility of human flourishing. Justice, from this perspective, does not merely concern the rights and resources a person has, it also concerns a person's ability to translate those rights and resources (e.g., income) into actual achievements, given the full range of circumstances the person faces.[9] For instance, if a person faces environmental harms that cause permanent neurological damage, then she faces a deprivation of her abilities to learn and reason, and to engage in activities and accomplishments that such faculties make possible. Because political protections or monetary compensation cannot fully recover or compensate for loss of these abilities, personal exposure to environmental harms can leave people unequal with respect to their potential to pursue a variety of activities making up a virtuous and flourishing life. The neurologically damaged person, for instance, will be unable to translate the provision of political rights and economic protections into the achievement of using her senses "to imagine, think, and reason—and to do these things in a 'truly human' way." Climate change, as I now discuss, can similarly undermine people's capabilities, putting the possibility of living a dignified human life beyond a person's reach.

Climate Change and Human Capabilities

Let us first consider some basic ways climate change will alter environmental conditions that constitute people's capabilities for *life* and *bodily health*. One frequently predicted consequence of rising global temperatures is climate variation that produces drought and further desertification in many areas on the African continent. The International Panel on Climate Change (IPCC) predicts that temperature change will produce declines in annual rainfall in various areas of South and North Africa (IPCC 2007, 237, 287). Additional changes in the onset of rain days and the variability of dry spells pose significant threats to agriculture, especially for small-scale farmers, who are expected to suffer the most severely. (448). In this context, even if a person living in this region of the world has control over an amount of land that can provide the food necessary for survival, and even if this person has the mechanical tools necessary for tilling her land, climate change may make her unable to translate these resources into the achievement of life and bodily health. For instance, because climate changes will increase drought, and because water is something without which crops needed for survival cannot grow, climate change will pose a direct threat to the capabilities of life and bodily health for at least some people.

In this case, climate change will also pose indirect threats to these same capabilities. For the landed farmer growing food for survival, drought will not only eliminate her opportunity to survive on her own, it is also likely to eliminate some of her opportunities for attaining the basic necessities for life and bodily health through alternative means (Stige et al. 2006; Mendelsohn et al. 2000). For example, if the landed farmer lacks water necessary for growing her own food, then larger, for-profit farmers probably will also lack water for growing crops; economic decline for these for-profit farmers may reduce the availability of employment options that might otherwise provide means of subsistence for the landed farmer. Likewise, if (as predicted) climate change also produces temperature changes that expand the normal geographical range and seasonal transmission period of disease-carrying insects, then it may limit the capacity of the landed farmer's community to provide for her basic needs, which otherwise might be expected under circumstances of temporary misfortune.

Other capabilities confront similar threats. Consider, for instance, the case of a climate exile, by which I mean a person displaced by climate change or climate-related environmental disasters, such as extreme flooding, or the submergence of one's home by the sea. In instances of climate-induced migration, where people are forced to leave not just their homes but also their home country, they will no longer be able to move freely from place to place as the *bodily integrity* capability requires, for currently there are no international laws that give asylum to climate refugees. (See Byravan and Rajan 2010, 242.) Likewise, a person facing conditions of such extreme legal instability and homelessness is unlikely to be able to pursue educational achievements or artistic activities requiring formal instruction, supplies, and a social context stable enough to engage one's creative capacities, as the *senses, imagination, and thought* capability would require.

Similarly, consider the capability for *control over one's environment*, which requires being able to participate in political choices that determine one's life and being able to hold property and seek employment. Not only can climate change undermine people's health in ways that make these achievements unattainable, a climate exile (even one who simply loses her home and land without being forced to leave her country) also is likely to face unknown periods of time in which none of these normal rights of citizenship are within her reach. (See Byravan and Rajan 2010, 242.) For instance, if one's life savings is in property, then submergence of this property may leave one with no real opportunity to hold property in the future. Likewise, the loss of one's home can disrupt one's material and psychological stability in ways that limit one's political voice. Such tragedy can also disrupt one's social networks in ways that diminish the viability of pursuing previously available employment options.

In each of these instances, climate change either directly or indirectly undermines the capabilities that make it possible for people to live what Nussbaum defines as a life that is worthy of the dignity of a human being—that is, a life in which the basic conditions of human flourishing are available. Before discussing the implications of these impacts for social institutions shaping the goals of ecological restoration, let us first consider how climate change might bear on one's *practical reason* capability, for this is the capability that has perhaps the least obvious relationship to climate change.

"Practical reason," the sixth capability on Nussbaum's list of ten central human capabilities, requires "being able to form a conception of the good, and to engage in critical reflection about the planning of one's life." In order to appreciate how climate change relates to this capability, consider what we might expect to be much less damaging effects of climate change on people in the United States, the world's richest country, which contains vast areas of habitable land that are above sea level. One consequence of climate change that will significantly influence people living in the United States is an increase in storm frequency and severity. Hurricane Katrina, which struck Louisiana's Gulf Coast in September of 2005 exemplifies the kind of weather event that a warming global climate will make more frequent. Indeed, some attributed the severity of Hurricane Katrina to warmer sea surface temperatures in the Gulf of Mexico, which is an expected consequence of climate change (IPCC 2007, 377).

This Category 3 hurricane had a devastating impact on residents of the entire Gulf Coast region, including citizens living in the city of New Orleans, Louisiana. In addition to destroying city residents' homes, those who fled the storm as well as those unable to flee experienced a level of fear, trauma, and dislocation that had profound psychological effects both during and after the weather event. A Harvard University study documenting post-evacuation responses of Hurricane Katrina's victims found that 86.4 percent of the study's respondents experienced significant financial, income, or housing loss as a consequence of the hurricane. Additionally, 36.3 percent experienced extreme physical adversity (e.g., sleeping on the floor, lack of food) and nearly 23 percent experienced extreme psychological adversity (e.g., having to go to the bathroom in front of other people, and threats of physical violence) (Hurricane Katrina Community Advisory Group 2006). Experiences such as these that occur in conjunction with natural disasters frequently produce significant mental health disorders among postdisaster survivors (Smith et al. 1990; Galea et al. 2002). More generally, complex events such as climate-related environmental disasters can affect how victims perceive and comprehend their lives, both immediately after an event and over time (Edelstein 1988, 9). One way to understand the implications of this for a person's practical reason capability is to consider how such events impact what David Edelstein refers to as a person's "lifescape" (11).

In documenting the impact of human exposures to environmental toxins, Edelstein defines "lifescape" as "our fundamental understanding about what to expect from the world around us." It is a reflection of our own unique and individual way of interpreting the world as well as our shared personal and social interpretive frameworks for understanding the world. A dramatic climate-related weather event can impact these frameworks in ways that have compelling implications for the practical reason capability.

Specifically, the practical reason capability requires being able to think about the planning of one's life in the context of complex forms of discourse, concern, and reciprocity with other people.[10] Traumatic lifescape changes can threaten this capability because they can shift the baseline assumptions one holds about one's self and one's relation to the surrounding world. For instance, the shift can involve moving from a baseline assumption that one has personal control over one's own actions and one's relation to the world one inhabits, to an assumption that one lacks personal control in both of these areas. Likewise, the shift can involve moving from an assumption that one's home is secure and valued, to experiencing the home as a threat to one's security and sense of self; it can involve a shift from having a sense of social trust to experiencing a sense of both social distrust and stigmatization (Edelstein 2004, 66). By altering one's baseline assumptions, these changes can easily undermine one's sense of individual agency, and, hence, the self-esteem necessary for forming one's own conception of the good and a plan of life for achieving it, which the practical reason capability requires.

In Western societies, for instance, part of a person's sense of personal control has to do with her belief that she can understand and exert control over her physical and social environment (Edelstein 2004, 89; Gibbs 1989; Janoff-Bulman 1985; Janoff-Bulman and Frieze 1983). Much like the impact of toxic exposures, catastrophic weather events such as Hurricane Katrina can easily break down one's belief in the world's predictability and controllability. These events are experienced as an involuntary intrusion on one's life, similar to the way a violent crime can involuntarily intrude upon one's life simply because one happens to be in the wrong place at the wrong time. If a person is consequently rendered helpless in protecting herself (and/or her family) from the resulting harms, then an experience of dependency on others may ensue, leaving the person unable to even make decisions influencing the various factors that determine her well-being.

This sequence of events characterizes what many victims of Hurricane Katrina suffered. Consider, for example, the loss of personal control as well as the threat to security and self experienced by hurricane victims housed in trailers built with materials that unknowingly exposed them to harmful levels of formaldehyde (Hsu 2008). Likewise, dependence on others for control over the factors affecting one's own well-being is descriptive of the situation faced by many residents of New Orleans' Ninth Ward, one of the poorest neighborhoods in the city. This neighborhood underwent infrastructural damage so extreme that many residents gave up their hope of ever returning.[11] More generally, according to the Hurricane Katrina Community Advisory Group's survey of hurricane victims, "Concern about the slow pace of infrastructure reconstruction was by far the most frequently mentioned consideration in the decision whether or not to return" (Hurricane Katrina Community Advisory Group 2006).

In these ways climate change will bring on weather events that undermine the practical reason capability. This capability enables people to exercise their human capacity as rational and thinking beings, that is, as more than just cogs in a machine to be shuttled along helplessly in the wake of climate-related disasters. Such disasters can alter what people do and how they think about themselves, their families, their communities, and the political and social relationships that make up their world. If these relationships are disrupted, if one's family and community is dispersed in different or unsecure places, and if one experiences no real possibilities for any agency in determining the opportunities and directions that one's life may take, then one will have neither the means nor the motivation to form one's own conception of what a good life involves or to engage in critical reflection about the ways and means to achieve a virtuous and flourishing life. In other words, climate change threatens to undermine the kind of practical reasoning that is unique to the human form of life—a life that Nussbaum characterizes as worthy of the dignity of the kind of being that is human (2000, 5).

While there are many additional ways in which climate change can disrupt environmental conditions enabling all the basic protections that the capabilities approach defines as central to social justice, my primary aim in the preceding discussion has been to establish that climate change can create changes in environmental conditions that threaten both materially basic and cognitively sophisticated human capabilities. For this reason, a stable climate system—by which I mean a climate system that changes at a rate to which humans can reasonably adapt without experiencing unnecessary threats to their capabilities—is an important part of the broader ecological context that supports the capabilities that Nussbaum seeks to protect as constitutional guarantees. Without accounting for this broader ecological context (and the role of a changing climate system that can radically disrupt it) as a precondition of these capabilities, Nussbaum's capabilities approach cannot ensure that people will have the environmental resources or experience the environmental contexts necessary for achieving social justice.[12] This is why a stable climate system should be understood as one component of a broader environmental meta-capability that supports and enables all of the central human capabilities on Nussbaum's list.

Elsewhere I have defined this environmental meta-capability as "Sustainable Ecological Capacity" (Holland 2008a). To have this meta-capability is to be able to live one's life in the context of ecological conditions that can provide environmental resources and services that enable the current generation's range of capabilities; to have these conditions now and in the future. Like all of the capabilities on Nussbaum's list, the environmental meta-capability is articulated in a general and abstract way to allow for its specification such that it can fit the distinct ecological circumstances of different communities (Nussbaum 2000, 77). For instance, while a stable climate system may be closely connected to people's capabilities for *life* and *bodily health* in sub-Saharan Africa, it may be closely connected to the *senses, imagination, and thought* capability of those who have a strong religious affiliation to a grove of ancient trees that climate change threatens to submerge. This variation in the kinds of ecological contexts that are necessary for protecting human capabilities is why climate policy should aim to prevent radical and rapid changes in both local ecosystems, as well as broader ecological systems, cycles, and processes that produce the resources and conditions enabling human capabilities.[13] Indeed, as the examples discussed previously show, the impact of climate variation on capabilities will sometimes occur through changes in weather systems (e.g., increases in the severity of storms); they will sometimes occur through changes in large-scale ecological cycles (e.g., groundwater depletion caused by drought); and they will sometimes occur through changes in the composition of ecosystems (e.g., onslaught of disease-carrying bugs in different geographical areas).

When Sustainable Ecological Capacity is understood to include a stable climate system, then the basic protections that a capabilities approach to justice requires will entail ensuring that changes in global climate occur at a rate to which humans can reasonably adapt; this means without experiencing significant threats to the capabilities that define the conditions of a minimally decent and dignified human life. As I now discuss, this characterization of what the possibility of human flourishing requires has specific implications for our current thinking about what it means to restore any particular ecosystem in an era of global climate change.

Human Capabilities and the Restoration of Ecosystems

If a stable climate system is understood as part of a broader environmental meta-capability enabling all the capabilities that government should protect, then what should be the goal of ecological restoration? As others in this volume elaborate, the philosophical debate in restoration ecology is delineated by two competing views. Early definitions of ecological restoration define "historic fidelity" as the goal. This entails returning degraded ecosystems to their predisturbance state, which is one that emulates the structure, function, diversity, and dynamics characterizing the ecosystem in a historical, less degraded condition. (See Throop, chapter 2, this volume; Higgs 2003, 107.) In some cases, this has meant returning the ecosystem's composition to the exact mix of species found in a previous time, and otherwise seeking to replicate the ecological relationships of the ecosystem in its former, more indigenous form.

Contemporary definitions of ecological restoration, in contrast, trend toward articulating goals that are achievable in the variety of contexts in which restoration can take place and that recognize the variety of meanings that restoring ecosystems can confer. Here the goals of restoration "should not try to move an ecosystem backward toward some idealized 'natural' structure" (Throop, this volume). Instead, today's leading voices in restoration ecology advocate goals ranging from some sense of loyalty to predisturbance conditions, to focusing entirely on the desired characteristics of the ecosystem in the future, and presumably, to what people in the present value as yet to come (for the former, see Higgs 2003, 127; for the latter, Hobbes and Harris 2001, 239). Thus, at the far end of this spectrum of contemporary restorationists are those that displace a commitment to historic fidelity with a commitment to what we might refer to as a "future desires" model of ecological restoration.

Between these two extreme perspectives are more modest contemporary views that maintain some loyalty to predisturbance conditions while not requiring the exact reproduction of the historic system's species, structure, and functioning. At first glance, this "modest view" seems like a practical compromise between (what is in many instances) an unrealistic commitment to an unachievable past ecosystem condition and a dangerous commitment to ecosystems that function so as to merely satisfy the whims we might predict will come from a highly consumptive and pleasure-seeking future population. However, while perhaps most practical, the modest view faces a significant problem: Although people are often moved by their connections to values and relationships of a past time and place, and will therefore infuse the modest view with a commitment to historic fidelity, ultimately these considerations "rest in human interests, [and] they may easily be trumped by other human interests that would violate historical fidelity" (Throop, this volume). Thus, the danger of the modest view of ecological restoration is that even if people in the present are committed to values made possible by restoring historic ecosystems, in the real political world where restoration takes place, values associated with the future desires model of ecological restoration may prevail.[14] For Throop, the modest view therefore runs the risk of divorcing restoration goals from any sense of "who we ought to be *in relation to the land*" (Throop, chapter 2, this volume).

Before considering how a capabilities approach to ecological restoration might inform this debate, first consider the different views of ecosystems these three goals of restoration imply. For restorationists committed to historic fidelity, ecosystems are providers of a valued past. Whether the aim is to produce the ecological resilience characterizing these systems or the human practices made possible by the historic form of their functioning, these systems in some way demonstrate a commitment to what people in the present value about the past. In contrast, for restorationists committed to a future desires model of ecological restoration, ecosystems are providers of a valued future. Whether the aim is to produce ecological systems that can provide resources to satisfy an increasing array of consumer goods or some other view of the future, these systems in some way demonstrate a commitment to what people will want at a future time. The third view of ecological restoration, which I have referred to as the modest view, treats the goals of restoration as adaptable to both past values and present views about what ecosystems should provide in the future. Here ecosystems are potentially providers of a richer set of human values than a highly consumptive future society might wish to protect, but they are not optimistically expected to function in a way that climate change is likely to make impossible, given that any specific historic species may be unable to thrive in a system subject to the predicted temperature changes.

Thus, the modest view is important because it offers a view of eco-systems that can avoid the moral bankruptcy of restoring ecosystems that merely satisfy goals emerging from highly consumptive lifestyles as well as the unrealistic goal of restoring ecosystems that cannot thrive in a changing global climate. Unfortunately, because the modest view of eco-logical restoration can easily give way to the moral bankruptcy that the future desires model allows, it should incorporate normative commit-ments that prevent an irreversible movement toward the more base human instincts driving efforts to restore ecosystems that merely function to satisfy highly consumptive lifestyles dominating today's consumer society. The modest view matters because even if these lifestyles were deemed morally acceptable, they are undoubtedly unsustainable for the planet-wide population. With this in mind, let us consider how a capa-bilities approach to ecological restoration can inform the goals of the modest view.

When understood to include an environmental meta-capability, the capabilities approach to justice requires protecting ten central human capabilities that make a dignified and virtuous life possible as well as the broader ecological conditions that are instrumental to those capabilities. Humans, in this context, are viewed as developmental beings, capable of flourishing in many different life activities. But such flourishing requires the protection of ecological conditions that enable capabilities, for without these conditions people will be unable to choose to pursue goals that the ten capabilities make possible. Additionally, such flourishing requires that these ecological conditions be sustained over time, so that the choice to live a fully dignified human life is available now and in the future; that is, no set of values dominating at one particular time should be able to squelch the option for diverse forms of human action that make up humans' evolving potential to develop in ways expressing their capacity to reason as rational, moral, social, and emotional beings.

Thus, while a capabilities approach to ecological restoration does not require that the modest view take a particular position on the issue of "who we ought to be in relation to the land," it does seek to sustain ecosystems that allow people to act toward nature in ways reflecting their own evolving views on the issue. This will mean dif-ferent things in the context of different communities, but crucially, it requires sustaining options for living a fully human life in relation to nature across the range of different communities that actually exist. Here, the goal of restoration is sustaining the option for living in relation to ecosystems that are presently available. Thus understood, ecosystems are the providers of present options to live a life that is true to the uniquely human form of living and, therefore, worthy of the dignity of a human being.

Global climate change poses an immediate obstacle to this goal, for temperature changes can easily wipe out a particular species to which a community has as felt attachment (e.g., maple trees), so that the experience of that community's attachment (e.g., making maple syrup together) is not available to future members of the community. While the obvious goal in that context is to restore an ecosystem that will make the loss of that attachment least burdensome—perhaps by introducing a climate-resilient species that allows for experiences similar to the one provided by maple trees that lack climate resilience—it is important to note that the cause of the problem for that particular (maple syrup making) community is the broader failure in other communities to restore degraded ecosystems in ways that promise to protect capabilities presently available. Specifically, the use of ecosystems by some communities for goals characterizing a future desires model of ecological restoration renders those systems incapable of protecting and sustaining the use of ecosystems in other communities for living a minimally decent and dignified life. In light of this tradeoff, an appropriate institutional response would redirect or otherwise constrain the activities of communities allowing degradation aligned with a future desires model of ecological restoration so that those ecosystems can perform the activities allowing other communities to interact with ecosystems in a manner protecting their central human capabilities.

Such a capabilities approach to ecological restoration adopts a view of *ecosystems as a sustainer* of present options over time, and it puts this understanding of ecosystems at the heart of a modest view of ecological restoration. The humans that ecosystems sustain are capable in the sense of being able to live a life that is worthy of human dignity. While this may mean having relationships with a historic species that reflects a particularly meaningful way of living in relation to the land, it may also mean enabling people to experience the pleasure of having luxury goods. However, because the consumption provision of those goods can easily degrade ecosystems in ways that eliminate the possibility for others to live a minimally decent and dignified life (e.g., by significantly contributing to climate change), the capabilities approach would prioritize the preservation of existing present options rather than allowing for the degradation of systems resulting from what some particular group of people think ecosystems should provide in the future.

Conclusion

This chapter has explored how a capabilities approach to social justice might inform the goals of ecological restoration in an era of global climate change. I started with Martha Nussbaum's capabilities approach, which grounds the conditions of social justice in human dignity and the activities and achievements possible in a life that is worthy of it. I then considered how climate change can undermine the ten capabilities that Nussbaum establishes as necessary for living the kind of life that is uniquely human in its form. Because climate change can undermine these capabilities, I argued that a stable climate system should be understood as part of a broader environmental meta-capability that enables all ten central human capabilities that Nussbaum seeks to protect as a matter of basic justice. Drawing on a capabilities approach that includes this meta-capability, I then considered the implications for ecological restoration. In the context of philosophical debate about ecological restoration, a capabilities approach would require restoring ecosystems that can sustain the conditions of human flourishing that are central to living a minimally decent and dignified life.[15]

Acknowledgments

For help with this chapter I thank the participants of the "Human Flourishing and Restoration in the Age of Global Warming" conference that took place in September 2008 at Clemson University. Allen Thompson and Jeremy Bendik-Keymer created an engaging intellectual environment at this conference and I appreciate their helpful comments throughout the review and editorial process. I benefitted, especially, from several conversations with Jeremy Bendik-Keymer, who taught me about ecological restoration and helped me think through its implications for human capabilities.

Notes

1. See Dale Jamieson, chapter 9, this volume, explaining that economic ways of reasoning dominate the contemporary policy debate to such an extent that the assumptions (of self-interested behavior) and biases (toward economic efficiency) of this perspective have become almost invisible.

2. Compare with Kawall, chapter 11, this volume, on greed.

3. See also Bendik-Keymer, chapter 13, this volume, and Vogel, chapter 15, this volume.

4. See Schlosberg, chapter 8, this volume, for a discussion of adapting society to protect the flourishing of nonhuman life in a world altered by climate change.

5. See Jamieson, chapter 9, this volume.

6. See Hirsch and Norton, chapter 16, this volume.

7. How the capabilities approach can address the interests of women is taken up most completely in Nussbaum 2000; the remaining issues are taken up most completely in Nussbaum 2006.

8. For present purposes, I will focus on human capabilities. However as Keulartz and Swart have just explored in chapter 6, Nussbaum (2006) has also considered what justice requires for nonhuman animals, assuming that animals are intrinsically valuable in the sense of having ends that society should promote. Similarly, Schlosberg (2007; chapter 8, this volume) has proposed expanding Nussbaum's theory of justice for animals to include the intrinsic value of other individuals and communities that make up nonhuman nature. While I am sympathetic to these arguments, I do not consider them here because my primary aim is to consider the kinds of protection that justice requires when we fully consider the instrumental value of the ecological systems and environmental resources that contribute to human capabilities.

9. For a discussion of the distinctions between rights and resources and capabilities, see Sen 1982, 357–358, and Sen 1992, 110.

10. For this characterization of the practical reason capability, see Nussbaum 2000, 82. Some of the lifestyle impacts of Hurricane Katrina are captured by the following findings of the Hurricane Katrina Community Advisory Group (2006): Eighty percent of the victims who were residing in a different county after fleeing the hurricane did not plan to move back to their pre-hurricane home; nearly 25 percent of the people who did return to their pre-hurricane home planned to move to a different county in the near future.

11. See Horne 2008, 212–214, for a detailed account of one family's response to the devastation of their home in New Orleans' Ninth Ward neighborhood.

12. See Holland 2008a for a fuller elaboration of this argument.

13. See Hirsh and Norton, chapter 16, this volume, arguing that a new metaphor is necessary for moving current policy thinking beyond the ecosystem scale to include these broader planetary processes. See Holland 2008b arguing why this goal of climate policy would require establishing capability ceilings in addition to the capability thresholds that Nussbaum already advances as necessary for social justice.

14. To take just one example, see Ernst 2003 for a discussion of the insurmountable political and economic barriers to successful restoration of the Chesapeake Bay.

15. Consider how this conclusion could be generalized across other approaches to ecosystem management including conservation policies.

References

Byravan, Sujatha, and Sudhir Chella Rajan. 2010. The Ethical Implications of Sea-Level Rise Due to Climate Change. *Ethics & International Affairs* 24 (3): 239–260.

Edelstein, Michael. 1988. *Contaminated Communities: The Social and Psychological Impacts of Residential Toxic Exposure*. Boulder, CO: Westview Press.

Edelstein, Michael. 2004. *Contaminated Communities: Coping with Residential Toxic Exposure*. 2nd ed. Cambridge, MA: Westview Press.

Ernst, Howard. 2003. *Chesapeake Bay Blues: Science, Politics, and the Struggle to Save the Bay*. Lanham, MD: Rowman & Littlefield.

Galea, Sandro, Jennifer Ahern, Heidi Resnick, Dean Kilpatrick, Michael Bucuvalas, Joel Gold, and David Vlahov. 2002. Psychological Sequelae of the September 11 Terrorist Attacks in New York City. *New England Journal of Medicine* 346 (13): 982–987.

Gibbs, Margaret. 1989. Factors in the Victim That Mediate between Disaster and Psychopathology: A Review. *Journal of Traumatic Stress* 2 (4): 489–511.

Higgs, Eric. 2003. *Nature by Design: People, Natural Process, and Ecological Design*. Cambridge, MA: The MIT Press.

Hobbes, Richard, and James Harris. 2001. Restoration Ecology: Repairing the Earth's Damaged Ecosystems in the New Millennium. *Restoration Ecology* 9 (2): 239–246.

Holland, Breena. 2008a. Justice and the Environment in Nussbaum's Capabilities Approach: Why Sustainable Ecological Capacity Is a Meta-Capability. *Political Research Quarterly* 61 (2): 319–332.

Holland, Breena. 2008b. Ecology and the Limits of Justice: Establishing "Capability Ceilings" in Nussbaum's Capabilities Approach. *Journal of Human Development and Capabilities* [Special Edition on Ideas Changing History] 9 (3): 399–423.

Horne, Jed. 2008. *Breach of Faith: Hurricane Katrina and the Near Death of a Great American City*. New York: Random House.

Hsu, Spencer. 2008. "Safety Lapses Raised Risks in Trailers for Katrina Victims." *The Washington Post*, Sunday, May 25. http://www.washingtonpost.com/wp-dyn/content/article/2008/05/24/AR2008052401973.html (accessed August 1, 2011).

Hurricane Katrina Community Advisory Group. 2006 *Katrina Baseline Report*, August 29, 2006. Harvard Medical School, Department of Health Care Policy, Cambridge, MA. http://www.hurricanekatrina.med.harvard.edu/baseline.php (accessed August 22, 2011).

Intergovernmental Panel on Climate Change (IPCC). 2007. *Climate Change 2007: Impacts, Adaptation, and Vulnerability (Contribution of Working Group II to the Fourth Assessment of the IPCC)*. New York: Cambridge University Press.

Janoff-Bulman, Ronnie. 1985. The Aftermath of Victimization: Rebuilding Shattered Assumptions. In *Trauma and Its Wake: The Study and Treatment of Post-Traumatic Stress Disorder*, ed. Charles Figley, 15–35. New York: Brunner/Mazel.

Janoff-Bulman, Ronnie, and Irene Hanson Frieze. 1983. A Theoretical Perspective for Understanding Reactions to Victimization. *Journal of Social Issues* 39: 1–17.

McMichael, Anthony, Diarmid Campbell-Lendrum, Carlos Corvalan, Kristie Ebi, Andrew Githeko, Joel Scheraga, and Alistair Woodward, eds. 2003. *Climate Change and Human Health: Risks and Responses*. Geneva: World Health Organization.

Mendelsohn, Robert, Wendy Morrison, Michael Schlesinger, and Natalia Andronova. 2000. Country-Specific Market Impacts from Climate Change. *Climatic Change* 45 (3–4): 553–569.

Millennium Assessment Panel. 2003. *Ecosystems and Human Well-Being: A Framework for Assessment*. Washington, DC: Island Press.

Nussbaum, Martha. 2000. *Women and Human Development: The Capabilities Approach*. New York: Cambridge University Press.

Nussbaum, Martha. 2006. *Frontiers of Justice: Disability, Nationality, Species Membership*. Cambridge, MA: Belknap Press of Harvard University Press.

Schlosberg, David. 2007. *Defining Environmental Justice: Theories, Movements, and Nature*. Oxford, UK: Oxford University Press.

Sen, Amartya. 1982. Equality of What? In *Choice, Welfare, and Measurement*, 353–369. Cambridge, MA: Harvard University Press.

Sen, Amartya. 1992. *Inequality Reexamined*. Cambridge, MA: Harvard University Press.

Smith, Elizabeth, Carol North, Robert McCool, and James Shea. 1990. Acute Postdisaster Psychiatric Disorders: Identification of Persons at Risk. *American Journal of Psychiatry* 147 (2): 202–206.

Stige, Leif, et al. 2006. The Effect of Climate Variation on Agro-Pastoral Production in Africa. *Proceedings of the National Academy of Sciences of the United States of America* 103: 3049–3053.

Thayer, Robert. 2003. *Lifeplace: Bioregional Thought and Practice*. Berkeley: University of California Press.

Tucker, Robert, ed. 1978. *The Marx-Engels Reader*. 2nd ed. New York: W. W. Norton & Company.

United Nations Environment Program (UNEP). 2007. *Global Environment Outlook (GEO-4)*. Hertforshire, UK: Earthprint.

Wackernagel, Mathis, et al. 2002. Tracking the Ecological Overshoot of the Human Economy. *Proceedings of the National Academy of Sciences of the United States of America* 99 (14): 9266–9271.

Worldwatch Institute. 2010. *State of the World: Transforming Cultures, from Consumption to Sustainability*. New York: W. W. Norton & Company.

8

Justice, Ecological Integrity, and Climate Change

David Schlosberg

The "restoration" of humanity—or, more directly, humanity's adaptation to a coming world of climate change—will come only with recognition of the human place within the rest of the natural world. In line with the themes of this volume, one key element of our adaptation to a changing climate will be a rethinking of ourselves as we interact with, and relate to, the nonhuman realm. Restoring humanity entails changing how we understand and how we relate to the lives and functioning of others—from individual animals to large-scale ecosystems—with which we are embedded in the world.

Political theorists face this task as clearly as the rest of the species. As Latour (2004, 58) has argued, political theory "abruptly finds itself confronted with the obligation to *internalize* the environment that it had viewed up to now as another world."[1] The argument here is that this type of internalization, confrontation, and restoration can be mediated through a broad conception of climate justice that encompasses not only human needs, but those of the natural world as well. A response to climate change that sees adaptation as a process of reconciling human and ecological integrity can be developed through use of the notion of capabilities. Such an approach gives us guidance for addressing two moments of adaptation: how to respond to both local and large-scale impacts on the natural world due to changes in climate; and how to adapt ourselves to fit with, and be part of, the natural world in which we function.

In this chapter, I first lay out a capabilities-based conception of justice, before examining how others have applied the approach to environmental issues. I then use the discourse of the environmental and climate justice movements to explore how capabilities are used in these movements, and are expanded to encompass a concern for communities as well as individuals. Finally, I lay out an argument for extending a capabilities-based approach to the protection of environmental systems.

Capabilities and Justice

Amartya Sen and Martha Nussbaum (Sen 1985, 1999a, 1999b; Nussbaum and Sen 1992; Nussbaum 2000, 2006) have developed an approach to thinking about justice that contrasts significantly with liberal theories that focus primarily on ideal forms or justifications for the distributions of goods. Their central argument is that we should judge just arrangements not simply in terms of the distribution of goods, but more specifically by how those distributions affect the ultimate well-being and functioning of people's lives. The central question of justice, for Sen, is about "the freedoms generated by commodities, rather than on the commodities seen on their own" (1999b, 74). For Nussbaum, the capabilities approach is not about how many resources a person commands, but instead what she is actually able to do and be: "We ask not just about the resources that are sitting around, but about how those do or do not go to work, enabling [her] to function in a fully human way" (Nussbaum 2000, 71). As chapters 6 and 7 in this volume have stated, our potential for functioning and flourishing is what is ethically significant, and injustice is found in the limitation of that potential.

Key to this argument is the concept of capabilities—those elements of a life that enable human beings to function. Rather than focus on the distribution of goods, Sen and Nussbaum argue that justice entails the ability to transform those goods into the potential for a functioning life; that ability is based in the availability of the capabilities necessary for such a transformation. The main task of a capability-based theory of justice, then, is to establish what is needed to transform primary goods into the potential for a fully functioning life.

Sen and Nussbaum offer differing accounts of a capabilities approach. Sen's original focus was on rethinking official quality of life assessments in order to improve development programs; his impact is seen in the *Human Development Reports* of the United Nations Development Programme (UNDP). The point of such an account of development is to focus on the underlying conditions necessary for people to have fully functioning lives, and to allow people to choose those lives for themselves. For Sen, a capabilities approach is less about perfecting an ideal theory, and more about solving key human problems (Sen 2009).

In addition, and in deference to democratic autonomy, Sen has refused to offer a predetermined list of specific capabilities; he only emphasizes a broad list of basic individual and social liberties and freedoms: political liberties, freedom of association, economic facilities, social opportunities, transparency guarantees, protective security, and a variety of economic and social rights (Sen 1999b). Public deliberation is the preferred method to develop more specific, contextual capability sets. As Sen argues, the "problem is not with listing important capabilities, but with insisting on one pre-determined canonical list of capabilities, chosen by theorists without any general social discussion or public reasoning. To have such a fixed list, emanating entirely from pure theory, is to deny the possibility of fruitful public participation on what should be included and why" (2005, 158). One of the necessary freedoms that people have is this opportunity to determine the capabilities necessary for functioning in their own communities. Political participation is key both as a capability in itself and as a tool for determining additional locally defined capabilities. This focus on participation and self-determination, as I will illustrate, has become a central issue in climate justice movements.

Nussbaum has rather less faith in such a democratic approach, and much more in legal and human rights frameworks; she claims the capabilities approach is "one species of human rights approach" (2006, 78). Nussbaum argues for a set of specific capabilities "as a foundation for basic political principles that should underwrite constitutional guarantees" (2000, 70–71). Focused on a conception of universal rights, Nussbaum has developed a detailed, basic "capability set" she sees as necessary for the functioning of any human life. See Holland, chapter 7, this volume, for the full list of capabilities in the set. Nussbaum argues that these capabilities, based in constitutional guarantees rather than community deliberation, would provide the social and political bases for the development of a fully functioning life (2006, 290–291).[2]

Either way we define the specifics of how we define capabilities, it is important to understand that the approach is not simply an alternative to other theories of justice. One of the central strengths of capabilities theory is that it can incorporate a number of issues of justice both within and, just as important, outside of the distributional paradigm. Recent arguments for the inclusion of social recognition as an element of justice, so forcefully developed by Young (1990), Fraser (1998, 2000), and Honneth (1995), can be brought under the capabilities umbrella. Nussbaum insists that the social bases of self-respect and nonhumiliation are a central capability—agreeing with theorists who argue for recognition as a central determinant of justice. Likewise, arguments for political participation and procedural justice—through public deliberation in Sen and "control over one's environment" in Nussbaum—are also encompassed in a thorough conception of the variety of capabilities necessary to construct a functioning life. Rather than an alternative to other theories of justice, then, we can understand a capabilities approach as more broadly encompassing many concerns of justice theorists.

One of the arguments here is that, theoretically, a capabilities approach encompasses the direction in which much of the discussion of climate justice has been moving. The approach addresses those things that are basic for human life—qualities that are necessary if we are to implement a notion of climate justice that allows for the basic functioning of human communities and the environment central to that functioning. Crucially, the capabilities approach broadens the focus of justice to include a range of issues not normally addressed in the theoretical literature on environmental justice, ecological justice, or climate justice (for an exception, see Schlosberg 2007). Many approaches to climate justice focus purely on equity (Jamieson 2001; Singer 2004), and even those that have development out of poverty as their central theme remain tied to a goods or income basis of judging justice (see EcoEquity 2008a, 2008b). Only the literature on environmental rights begins to touch on the importance of a stable climate system as central to a functioning life (Caney 2005 and 2006; Vanderheiden 2008a and 2008b). But a capabilities approach to climate justice could also address, for example, specific ways the environment supports human capabilities, the diversity of necessary environmental capabilities depending on place, the political structure of decision making on climate mitigation and adaptation, and—I argue here—the capabilities necessary for nature itself to function.

Capabilities and the Natural World

To start, I want to discuss the different ways that a capabilities approach has been applied to the natural world. Both Sen and Nussbaum have addressed the issue of nonhuman nature, and have offered different ways to incorporate this realm in a capabilities approach to justice. Sen's contribution has been more limited, but addresses the question of the environmental bases of existing capabilities. In a discussion of various circumstances under which people could have the same level of income, and yet not be equal in terms of their actual well-being, Sen notes that "variations in environmental conditions, such as climatic circumstances (temperature ranges, rainfall, flooding and so on), can influence what a person gets out of a given level of income" (1999b, 70). Here, Sen wants to pay attention not just to the resources we get, but also to how environmental circumstances can have a serious impact on our ability to construct functioning lives.

Sen has also commented on the question of the relationship between capabilities and nature with a focus on providing sustainability for future generations. Anand and Sen (2000, 2035) argue that "we can talk of sustainability only in terms of conserving a capacity to produce well-being" for people in the future. "The moral obligation underlying sustainability is an injunction to preserve the capacity for future people to be as well off as we are" (2038). Likewise, in discussing the preservation of endangered animals, Sen (2004, 1) focuses on the preservation, and possible expansion, of "the substantive freedoms of people today without compromising the ability of future generations to have similar, or more, freedoms." Sen's focus here is on the obligation we are under to extend capabilities that we currently have to others in both our own generation and future generations. He argues for the importance of future generations of humans to have the freedom to enjoy the same environmental benefits—from clean air to rare species—that earlier generations enjoyed.[3]

So Sen gives us two sets of reasons for incorporating attention to environmental conditions and systems into a conception of climate justice based on the capabilities approach. In the first, climatic variations impact what individuals are able to do with the resources that they have. If, in fact, climatic change makes it more difficult to grow food, or if climate-induced flooding uproots us from our homes, then climate change itself limits our capability to convert resources into functioning lives. It is climate change that can be seen as a barrier to that functioning, and so is a condition that is an injustice to us—or, more specifically, to those human beings that face this threat. Secondly, while we must pay attention to our obligation not to impose barriers on our contemporaries, we also have that same obligation to future generations of human beings. Any system that would limit the freedoms and capabilities of future generations, and limits the environmental possibilities for those generations, would be unjust. Any conception of a capabilities-based notion of climate justice must focus on the way that changes in the climate system, and related ecological systems, will impact the capabilities of other human beings, now and into the future.

Nussbaum's discussion of the capabilities approach and the natural world is a bit more nuanced—and extends beyond human beings. As Keulartz and Swart (chapter 6, this volume) explore, Nussbaum (2004 and 2006) has directly addressed the application of capabilities theory to nonhuman nature, in particular to some individual animals. She breaks crucial ground by arguing that a capabilities approach is applicable to a wide range of types of animal sentience and dignity, and of corresponding needs for the flourishing of these creatures. It is plausible, she claims, to think that our various relationships with nonhuman animals ought to be regulated by a conception of justice. Nussbaum argues that the capabilities approach "is capable of yielding norms of interspecies justice that are subtle and yet demanding, involving fundamental entitlements for creatures of different types" (2006, 327). Basically, Nussbaum's argument is that the capabilities list developed for human populations can also be applied to sentient animals, as a way of extending justice across the species boundary. Nussbaum's contribution here is significant and should not be understated; she offers a direct extension of the concept of justice beyond human beings alone.[4] She adds to a literature on doing justice to nonhuman nature that focuses on what it is we share with other parts of nature, rather than the qualities that differentiate us.[5] As valuable as it is, however, Nussbaum's approach to nature, and so the larger issue of ecological integrity, is limited in two key ways; I will return to these critiques shortly. First, however, I wish to explore how Nussbaum's approach can be applied specifically to climate justice.

Breena Holland and Ed Page have both suggested applications of Nussbaum's capabilities theory to climate justice. Both suggest the development of a new, environmental capability. Page advocates the adoption of a "safe and hospitable environment as a vital ingredient of a decent life rather than a facilitator of the other functionings." Specifically, Page suggests the adoption of the "capability to experience life in an environment devoid of dangerous environmental impacts such as those associated with climate change" (2007, 464).

For Holland (2008), however, simply *adding* a capability misses an important realization. She argues that without an accounting of the importance of functioning ecological systems, and how they enable all of the other capabilities, Nussbaum simply fails to identify what is actually necessary in order to achieve justice (Holland 2008, 6). Holland recommends a "meta-capability" that, in essence, enables all of the others. This new capability, which she labels "Sustainable Ecological Capacity," would involve being able *"to live one's life in the context of ecological conditions that can provide environmental resources and services that enable . . . capabilities."* Holland offers an argument for an "environmental justice threshold": the level at which ecological systems could sustain not only themselves, but also the other basic capabilities for human beings in the system. "As long as ecological systems have the functional capacity to sustain the conditions enabling the minimum threshold level of Nussbaum's capabilities for each person, the *ecological conditions of justice are met*" (10). Holland explicitly links this new capability to an understanding of climate change, both in previous work (Holland 2008) and in this volume.

Both Page and Holland offer reasonable extensions of capability theory into a conception of climate justice. If the capabilities approach is about flourishing, and we all flourish in particular environments, flourishing for human beings means providing for those ecological support systems that make our functioning possible. These capabilities can be seen as the precursors of constitutional principles and protections—Nussbaum's preferred approach.[6] And if capabilities are to be negotiable and subject to citizen deliberation, as Sen argues, then it may be the case that the level of public discourse on the impacts of climate change would justify such a new capability. Most important, in a world in which the discussion of climate change is quickly shifting from prevention and mitigation to adaptation, capabilities offer an important, implementable framework for response. In discussing adaptation, the climate justice discussion needs to incorporate recognition of the importance of survival and functioning, and to the role natural systems play in that functioning. Such a model would begin to address what exactly is needed—in terms of environmental and developmental conditions—to survive, function, and develop as human beings. The focus would be on how climate change makes our lives more vulnerable in terms of the impacts on the environmental basis of capabilities, and how a notion of climate justice can most directly address that vulnerability as we adapt to new environmental conditions.

The Public Discourse of Climate Justice Movements

As a way to ground and offer real-world support to the main point that a capabilities approach can be used to extend a notion of climate justice, it is important to note that the discourse of climate justice movements already includes and illustrates a concern with a wide range of capabilities as well as the concept of functioning.

Numerous statements of climate justice movements focus on the fact that global climate change will impact everyday life, and reduce "peoples' ability to sustain themselves" (Miller and Sisco 2002, 1). The Bali Principles of Climate Justice directly link the question of climate change to the ability of local communities to sustain their ways of life. Activists in the movement declare that the key to climate justice is protecting vulnerable communities, and what it takes for them to function. For example, the Indigenous Environmental Network is documenting in detail the impacts of climate change on traditional ways of life and indigenous communities' ability to sustain and reproduce their cultures. Activists in New Orleans after Katrina have focused on similar themes (Bullard and Wright 2009).

There are a variety of very specific capabilities that movement groups identify as central to the definition of climate justice. One clear example is the focus on community health. As the International Climate Justice Network puts it, the impacts of climate change will "threaten the health of communities around the world" (2002). The health impacts of climate change will most impact vulnerable communities. The issues include the effects of increased heat in many cities, the possible increase in severe weather events, increased diseases, decreased food security, and mental health impacts. Environmental refugees, impacted by any number of these issues, will be a further burden on community functioning—both in their home communities and those they are forced to move to.

Ecological integrity is another capability that the movement addresses. New Orleans serves as an interesting example here. Before Katrina, environmental justice organizing in Southern Louisiana focused on "cancer alley," the oil refineries and petrochemical factories that activists accused of threatening public health. In the aftermath of Hurricane Katrina there is a new awareness and understanding of how of the climatic effects of such carbon-intensive industries rebounded to threaten the very survival of the city in the form of the devastating storm, as well as how long-term neglect of the area wetlands' physical integrity made Katrina's impact even worse.

The vulnerability of indigenous nations and small island states is already central to activist notions of climate justice. The argument is not only that some are more vulnerable than others in terms of the coming impacts of climate change; it is also that this vulnerability is not recognized or respected by the larger industrialized nations. Fear of a cultural diaspora—as happened in poor African-American neighborhoods in New Orleans post-Katrina—is also a major concern. Both of these considerations embody Nussbaum's conceptions of social recognition and affiliation as central capabilities, necessary for human functioning and for justice.

Related demands for authentic and broad public participation in the development of local and global climate policy are clearly put forth in numerous nongovernmental organization (NGO) demands for, and principles of, climate justice. All movement-based sets of principles on climate justice make clear that public and community participation should be accountable, authentic, and effective at every level of decision making.[7] Community participation is one of the ten key principles of the Environmental Justice and Climate Change Initiative: "At all levels and in all realms, people must have a say in the decisions that affect their lives. Decision makers must include communities in the policy process" (EJCC 2002). Activists in the climate justice movement declare that the key to climate justice and the protection of vulnerable communities is the expansion of democratic participation—a direct call for participation as a key capability.

One observation is crucial here. It is clear that the major concerns of climate justice groups are not just focused on individual capabilities, but also on community capability and functioning. This is not simply a concern for what a group or community provides for individuals in that community, but is also about the functioning of communities themselves. For movement groups, environmental injustice is seen as a process that takes away the ability of individuals *and their communities* to fully function, through poor health, destruction of economic and cultural livelihoods, and general and widespread environmental threats. It is the *community* functioning that is at issue *as much as*, and *distinct from*, individual functioning. These capabilities—whether bodily health, affiliation, control over one's environment (including political participation), a safe and hospitable environment, or sustainable environmental capacity—help not only individuals to function and flourish, but their communities as well. Many indigenous communities, for example, are impacted not only by threats to their individual capabilities, but also by events and practices that threaten their ability to reproduce cultural and spiritual underpinnings of their ways of life.[8] Community-based arguments for climate justice address the threats to numerous capabilities at both levels, and the particular desire for the continued functioning of communities in the face of climate change. If, as Bendik-Keymer suggests in his chapter in this volume, one of the tasks we face is to focus on a "virtue in context" or social approach, rather than just an individual one, these movements for community functioning illustrate that many groups have already embraced this challenge.[9]

Overall, the point here is that a major focus of climate justice movements is on the capabilities necessary for individuals and communities to fully function. Climate justice advocates exemplify and clarify the range of concerns central to a capabilities-based notion of justice—concerns for an environment that will support the functioning and flourishing of vulnerable human communities.

Capabilities, Climate Change, and Ecological Integrity

Building on models regarding capabilities and the natural world, as well as discourse of climate justice movements, I want to argue that it is possible extend the capabilities approach beyond its reference to humans alone, and into a consideration of the capabilities necessary for the nonhuman natural world to function. The argument here is one that applies capabilities to nonhuman animals and ecosystems without the necessary reference to them as only human support systems. Here I wish to explore two of the issues that are problematic, and limiting, in Nussbaum's extension of capabilities framework to the nonhuman realm. The first is Nussbaum's use of the experience of dignity as the central indicator of whether or not one deserves justice; integrity, I argue, is a much more applicable and apt indicator here. The second problem with Nussbaum's argument—but also with the capabilities approach generally as it has been applied to the natural world—is its individualist focus. Here the point is to explore a more systems- or community-based application of a capabilities approach. The broader argument is that we can take capabilities quite far in developing a framework for a much more extensive and, importantly, applicable theory of climate justice.

In justifying the extension of a theory of justice beyond human beings, Nussbaum focuses on one quality that, she argues, we share with other sentient animals—dignity. Humans and other sentient animals have lives worthy of respect, she insists, and it is an injustice if such an animal does not have the opportunity to develop, flourish, and lead a life with dignity. While Nussbaum does not directly equate human and animal dignity, she argues that "there is no respectable way to deny the equal dignity of species across species" (2006, 383). Following a capabilities approach, any life that hopes to attain such a dignified existence, argues Nussbaum, "would seem at least to include the following, adequate opportunities for nutrition and physical activity; freedom from pain, squalor, and cruelty; freedom to act in ways that are characteristic of the species. . . ; freedom from fear and opportunities for rewarding interactions with other creatures of the same species, and of different species; a chance to enjoy the light and air in tranquility" (326).[10] Dignity, then, requires similar capabilities for all sentient animals—human or otherwise. It is

Nussbaum's entre into extending the concept and practice of justice beyond the human realm.

While recognizing this important and paradigm-shifting move on Nussbaum's part, dignity seems a limited and problematic concept when applied beyond the human realm. Nussbaum herself refers to the type of dignity at the heart of the human rights tradition—one that basically signals unconditional *standing*. But in more general terms, including contemporary rights arguments, the term "dignity" is primarily understood as an individual psychological state referring to one's own self-respect.[11] It would seem an incredibly difficult task to draw the line, from one species to the next, as to whether or not this type of dignity is an achieved quality; it also seems quite difficult to equate what we, as humans, experience as dignity with what a dog, or lion, might understand (if, indeed, it could). While Nussbaum's basis for dignity is a general notion of standing, she refers to "a life with the type of dignity relevant to that species" (2006, 351), which seems to move away from a notion of universal standing toward a more variable idea.[12]

When it comes to the nonhuman realm, a focus on integrity would be a better choice than dignity; it applies more uniformly across a greater spectrum of nonhuman nature and refers more directly to the status of beings rather than any hint of psychology. Integrity can encompass the very straightforward understanding of physical or bodily integrity in the legal sense, including the basic right against violation of the body (either by other citizens or the state). The integrity of a being is a better marker for standing—the rights-based notion of dignity Nussbaum wants to support—than is psychological dignity (how people understand "dignity" today). An affront to what we may call the dignity of an animal itself may cause no clear injury, yet a violation of bodily integrity not only immediately threatens that animal, but also its long-term potential flourishing and that of others dependent on it in a natural system. Integrity, in this sense, represents the idea of noninterruption of functioning, which is, ultimately, what Nussbaum sees as the essence of justice. This shift from dignity to integrity does not necessarily change the focus of Nussbaum's argument; all of the capabilities in her ultimate list directly apply to the integrity of a life, not just dignity. In addition, "integrity" is a term that is applicable if we are to broaden our concern from individual animals to encompass larger natural systems. This leads to the second point regarding the shortcomings in Nussbaum's approach.

Nussbaum's extension of justice only applies to (some) individual animals, and not to other entities—either individual or collective—of the natural world. This illustrates that Nussbaum is focused on animal rights rather than a broader notion of justice that can be applied to ecological and climate systems. While pioneering in her move away from the limited possibilities of utilitarian and social contract approaches to animals, Nussbaum remains tied to a very individualist conception of liberal rights, a focus that unnecessarily limits the reach of a capabilities approach into environmental considerations, where individual potential—human or nonhuman—is nested in a functioning ecosystem.

Nussbaum is concerned with the flourishing potential of individual animals, and while she recognizes the importance of habitat, she justifies the limited focus on isolated animals by insisting "damage to the species occurs through damage to individuals" (2006, 357). This ignores the integrated reality of individuals within ecosystems, where the potential to function is clearly determined by a broadly functioning ecological system in which both humans and nonhumans are embedded and dependent. While damage to individuals can indeed decimate a species (as in overfishing), more often than not the source of extinctions is a loss of habitat and ecological support systems, including the various symbiotic relationships that are the heart of ecosystems. The functioning of communities and systems is absolutely crucial to the functioning of the individuals within them. Simply put, all human and nonhuman animals need a working environment to function; this environment includes not just other individual animals, but also nonsentient life and the relationships and interactions that sustain ecosystems. As Keulartz and Swart (chapter 6, this volume) state, the point is to focus both on animals and the habitats that support their functioning—human and nonhuman alike.

Ecosystems also have integrity, and definitions of ecological integrity are quite similar to those for human bodily integrity, focusing on the natural conditions necessary for ecosystems to continue to function and evolve (cf. Pimentel, Westra, and Noss 2000). At the base of Nussbaum's conception of justice is the Aristotelian notion of teleology—that injustice is based on interrupting the potential unfolding of the ends or purpose of a human being (or sentient animal). And yet we can also claim something like a purpose or particular type of flourishing in the way that ecosystems evolve. This is not to say that all ecosystems have a singular and universal end—ecologists are increasingly hesitant about any notion of ecosystem end-states or stability, and more open to a plurality of states and constant evolution. Like human beings, there is no single teleology for nonhumans or ecosystems, but rather a range of potential states that depend on the conditions and inputs to the individual or system. A notion of justice with regard to ecosystems would focus on how the teleological potential of individuals and systems, human and nonhuman, is undermined or interrupted by human actions and abuses.

It is the threat to the integrity and potential functioning of a variety of ecological systems that is at the heart of a problem like climate change. This is, unfortunately, the crucial essence of being that we currently share with other parts of nature due to climate change—the vulnerability of the integrity of our individual and community bodies, human and nonhuman. Carbon dioxide and other greenhouse gas emissions threaten the potential of a whole range of teleological entities—not just individual human beings. The challenge, as laid out throughout this volume, is a conception and practice of restoration that addresses our responsibility to restore functioning to both ourselves and impacted, undermined ecosystems.

My argument here is that the capabilities approach could enrich conceptions of climate justice by bringing recognition to the functioning of natural systems, as well as the human and nonhuman individuals within it. In this approach, the central issue of climate justice is the interruption of the capabilities and functioning of living systems—what keeps those living systems from transforming primary goods into the potential functioning, integrity, and flourishing of both human and nonhuman individuals within them. If injustice comes with the interruption of the capabilities necessary to function, we need to examine those interruptions—and, so injustices—at both the individual and community/system level.

In other words, a capabilities approach could be extended beyond what Holland offers in her chapter—attention to the way that human capabilities are dependent on natural systems—to one that also focuses on the fact that those natural relationships and processes also provide the potential for the functioning of other nonhuman flora, fauna, and systems. If we remain tied to a focus on what ecological systems do only for human survival and flourishing, we do not get at the central issue of why human beings are undermining the ability of various ecological systems to function; it does not help us get past our unwillingness to recognize the integrity of ecological systems not only for what they provide for *us*, but also for the work that functioning ecosystems do for all that is *not* human as well. A capabilities-based approach to restoration can support an ethical landscape that includes, and yet expands upon, the support of human functioning alone. When we interrupt, corrupt, or defile the potential functioning of ecological support systems, we do an injustice not only to human beings, but also to all of those nonhumans that depend on the integrity of the system for their own functioning.

There is no theoretical need to limit a capabilities-based approach to climate justice to individual human beings, and there are ecological reasons for extending it to systems as well as individual animals. It is the disruption and increasing vulnerability of the integrity of ecosystems that is at the heart of the injustice of climate change, both in terms of its impact on vulnerable human communities and nonhuman nature. A systems-based approach to capabilities theory helps us to flesh out what, exactly, the moment of injustice is in this principle, that point where human beings, nonhuman animals, and natural systems are denied the capabilities necessary to function. The capabilities approach offers an opportunity to determine which necessary capabilities are being undermined or disrupted. With this tool, we can identify specific vulnerabilities, and how those vulnerabilities differ from place to place, system to system. In understanding the injustice with regard to both human beings and the nonhuman world in this way, we can then turn our focus to the social, cultural, industrial, political, and institutional processes that deny necessary capabilities and so prevent potential functioning from being realized.

Conflicts and Conclusions

All notions of climate justice bring with them potential conflicts; most of these to date have been between developed and developing nations. A focus on historical responsibility for climate change brings resistance from developed nations who do not want to consider (or pay) the costs of their past actions. Emphasizing an equitable approach, with every individual having an equal right to a certain amount of carbon emissions, faces critiques from developing nations that fear their economic future being hampered.

Clearly, implementing an idea of ecosystem-based climate justice would have its own obstacles, though of a different type.[13] Potential conflicts are sure to come with attempts to broaden a capabilities approach to a system-wide focus with or over individuals. Conflicts could include individual human vs. human community, individual animal vs. ecosystem, individual humans vs. ecosystems, and human communities vs. ecosystems. Even with a focus solely on the functioning of human communities, there certainly could be a problem with elevating community functioning above the individual; such a view could lead to abuses of individuals and their rights.[14] We can anticipate the same types of conflict erupting between protectors of individual animals vs. protectors of ecosystems. Cripps has argued that it is this range of inherent conflicts that might make a capabilities-based notion of ecological justice "doomed from the outset" (2010, 14). And yet while it is true that in ecosystems there is much more "sacrifice" of individuals for the larger stability of the system—in particular in terms of animals who become food for others—it can be argued that to be such food is a form of functioning (see Schlosberg 2007; Hailwood 2011). Ultimately, each animal—human or non—takes from the natural world in order to survive, and then returns to offer sustenance to that same world. The difficulty, of course, is finding a balance that allows us to take from our surroundings without limiting the ability of others to do the same, and without limiting the ability of the larger systems to function. Both the conflicts between human and nonhumans, and conflicts between nonhumans themselves, are part of the set of natural relationships a theory of ecological justice strives to recognize and maintain.

While the specifics of engaging such potential conflicts must be further developed, they lie beyond the confines of this chapter. The argument here is that such conflicts can be addressed in various ways consistent with a capabilities approach to justice. In response to these potential conflicts, it seems feasible that a combination of basic individual rights as championed by Nussbaum, plus the type of public deliberation that Sen defends as a basic capability, should be invoked in the public discussion of the potential conflicts between the capabilities of individuals and natural systems. A thorough form of ecological reflexivity can assist us in avoiding the subjugation or use of individuals to serve the flourishing of the larger community.[15] Such an approach allows for difference in the way we define needs by locality; local participation and deliberation can help us to understand and determine the distinct and local environmental needs of various communities. Here, a capabilities-based approach, using deliberative tools, can allow for the type of participation, flexibility, and local knowledge that is crucial to understand and support both individual and systems functioning while meeting the participatory demands of both the capabilities approach and climate justice movements. Such an approach, crucially, is also an apt way of dealing with the major responses we will have to undertake as we adapt to a coming world changed by our alteration of the climate system.

That is really the essence of the current challenge, and why these issues should be addressed in a political framework of justice rather than in the more general, and frankly tired, argument over the intrinsic value of animals or nature. A version of a capabilities approach, based in systems and fleshed out by ecological reflexivity, can help us adapt to climate change in a just manner. The key shift, of course, is one in line with the themes of this volume: we must adapt not only our behavior to climate change, but also our understanding of, and ongoing reflexivity about, our place in relation to the natural world in which we function and flourish.

Notes

1. Compare Thompson, chapter 10, this volume, and Vogel, chapter 15, this volume, on internalizing the environment.

2. Tellingly, the index to *Frontiers of Justice* has no listing for "democracy," but an extensive one for "human rights."

3. This approach closely mirrors deShalit's (1995) argument for a version of communitarian intergenerational justice, which includes a duty to preserve the natural conditions necessary for the survival of future generations.

4. There are a number of critiques of the implications of Nussbaum's approach; see, for example, Cripps 2010; Hailwood 2011; Ilea 2008; Schlosberg 2007, chap. 6; and Wissenburg 2011.

5. See also, for example, Baxter 2005 and Low and Gleeson 1998.

6. This suggestion for providing capabilities as legal rights would fit with the arguments of those, such as Vanderheiden (2008a, 2008b), who base climate justice in environmental rights.

7. See, for example, the 10 Principles for Just Climate Change Policies in the U.S. (Environmental Justice and Climate Change Initiative 2002) and the Bali Principles of Climate Justice (International Climate Justice Network 2002).

8. For a further explication of this argument, see Schlosberg and Carruthers 2010, which examines community functioning in indigenous environmental justice battles.

9. See Vogel, chapter 15, this volume.

10. Nussbaum feels that this insistence on animal dignity may cross a liberal line into a notion of the good, so leaves it as "a metaphysical question on which citizens may hold different positions while accepting the basic substantive claims about animal entitlement" (2006, 383). The fallback is a less sweeping idea, one nonetheless still based in the recognition of the importance of animals in an overall theory of justice.

11. See, for example, Honneth's discussion of recognition and dignity, and Fraser's critique of the implications of such a victim-centered approach, in

Fraser and Honneth 2003. Such a critique of individual psychological states is the basis of Fraser's (1998, 2000) argument for a status-based, rather than psychological, notion of recognition (in human communities). While Fraser does not address Nussbaum, her focus on status clearly refers to the type of universal rights-based dignity that Nussbaum aims for.

12. She roots the eighteenth-century French notion of being *digne*—of deserving standing—in an Aristotelian understanding of an animal's proper teleology when healthy.

13. The first being the alienation Vogel, chapter 15, this volume, conceptualizes.

14. For example, foot binding and genital mutilation for the sake of culture and tradition.

15. For more on such a reflexive and deliberative approach see Schlosberg 2007, especially chap. 8.

References

Anand, Sudhir, and Amartya Sen. 2000. Human Development and Economic Sustainability. *World Development* 28 (12): 2029–2049.

Baxter, Brian. 2005. *A Theory of Ecological Justice*. London: Routledge.

Bullard, Robert, and Beverly Wright, eds. 2009. *Race, Place, and Environmental Justice After Katrina*. Boulder, CO: Westview Press.

Caney, Simon. 2005. Cosmopolitan Justice, Responsibility, and Global Climate Change. *Leiden Journal of International Law* 18: 747–775.

Caney, Simon. 2006. Cosmopolitan Justice, Rights and Global Climate Change. *Canadian Journal of Law and Jurisprudence* 19 (2): 255–278.

Cripps, Elizabeth. 2010. Saving the Polar Bear, Saving the World: Can the Capabilities Approach Do Justice to Humans, Animals and Ecosystems? *Res Publica* 16 (1): 1–22.

deShalit, Avner. 1995. *Why Posterity Matters*. London: Routledge.

EcoEquity. 2008a. The Right to Develop in a Climate Constrained World: The Greenhouse Development Rights Framework. 2nd ed., executive summary. http://gdrights.org/wp-content/uploads/2009/01/gdrs_longexecsummary.pdf (accessed November 1, 2011).

EcoEquity. 2008b. Greenhouse Development Rights. http://gdrights.org/wp -content/uploads/2009/01/thegdrsframework.pdf (accessed November 1, 2011).

Environmental Justice and Climate Change Initiative (EJCC). 2002. 10 Principles for Just Climate Change Policies in the U.S. http://www.ejnet.org/ej/ climatejustice.pdf (accessed March 25, 2011).

Fraser, Nancy. 1998. Social Justice in the Age of Identity Politics: Redistribution, Recognition, and Participation. In *The Tanner Lectures on Human Values*, vol. 19, ed. Grethe B. Peterson, 1–67. Salt Lake City: University of Utah Press.

Fraser, Nancy. 2000. Rethinking Recognition. *New Left Review* 3 (May/June): 107–120.

Fraser, Nancy, and Axel Honneth. 2003. *Redistribution or Recognition: A Political-Philosophical Exchange*. London: Verso.

Hailwood, Simon. 2011. Bewildering Nussbaum: Capability Justice and Predation. *Journal of Political Philosophy* 19 (1).

Holland, Breena. 2008. Justice and the Environment in Nussbaum's "Capabilities Approach": Why Sustainable Ecological Capacity Is a Meta-capability. *Political Research Quarterly* 61 (2): 319–332.

Honneth, Axel. 1995. *The Struggle for Recognition: The Moral Grammar of Social Conflicts*. Cambridge, MA: MIT Press.

Ilea, Ramona. 2008. Nussbaum's Capabilities Approach and Nonhuman Animals: Theory and Public Policy. *Journal of Social Philosophy* 39 (4): 547–563.

International Climate Justice Network (ICJN). 2002. Bali Principles of Climate Justice. http://www.ejnet.org/ej/bali.pdf (accessed March 25, 2011).

Jamieson, Dale. 2001. Climate Change and Global Environmental Justice. In *Changing the Atmosphere: Expert Knowledge and Environmental Governance*, ed. Clark A. Miller and Paul N. Edwards, 287–308. Cambridge, MA: MIT Press.

Latour, Bruno. 2004. *Politics of Nature: How to Bring the Sciences into Democracy*. Cambridge, MA: Harvard University Press.

Low, Nicholas, and Brendan Gleeson. 1998. *Justice, Society and Nature: An Exploration of Political Ecology*. London: Routledge.

Miller, Ansje, and Cody Sisco. 2002. "Ten Actions of Climate Justice Policies." *Second National People of Color Environmental Leadership Summit—Summit II Resource Paper Series*. Environmental Justice and Climate Change Initiative. http://www.ejrc.cau.edu/summit2/SummIIClimateJustice%20.pdf (accessed March 25, 2011).

Nussbaum, Martha C. 2000. *Women and Human Development: The Capabilities Approach*. Oxford, UK: Oxford University Press.

Nussbaum, Martha C. 2004. Beyond "Compassion and Humanity": Justice for Nonhuman Animals. In *Animal Rights, Current Debates and New Directions*, ed. Cass R. Sunstein and Martha C. Nussbaum, 299–320. Oxford, UK: Oxford University Press.

Nussbaum, Martha C. 2006. *Frontiers of Justice: Disability, Nationality, Species Membership*. Cambridge, MA: Harvard University Press.

Nussbaum, Martha C., and Amartya Sen. 1992. *The Quality of Life*. Oxford, UK: Oxford University Press.

Page, Edward. 2007. Intergenerational Justice of What, Welfare, Resources, or Capabilities? *Environmental Politics* 16 (3): 453–469.

Pimentel, David, Laura Westra, and Reed Noss. 2000. *Ecological Integrity: Integrating Environment, Conservation, and Health*. Washington, DC: Island Press.

Schlosberg, David. 2007. *Defining Environmental Justice*. Oxford, UK: Oxford University Press.

Schlosberg, David, and David Carruthers. 2010. Indigenous Struggles, Environmental Justice, and Community Capabilities. *Global Environmental Politics* 10 (4): 12–35.

Sen, Amartya. 1985. Well-Being, Agency and Freedom: The Dewey Lectures 1984. *Journal of Philosophy* 82 (4): 169–221.

Sen, Amartya. 1999a. *Commodities and Capabilities*. Oxford, UK: Oxford University Press.

Sen, Amartya. 1999b. *Development as Freedom*. New York: Anchor.

Sen, Amartya. 2004. Why We Should Preserve the Spotted Owl. *London Review of Books* 26 (3) (February 5). http://www.lrb.co.uk/v26/n03/amartya-sen/why-we-should-preserve-the-spotted-owl (accessed March 25, 2011).

Sen, Amartya. 2005. Human Rights and Capabilities. *Journal of Human Development* 6 (2): 151–166.

Sen, Amartya. 2009. *The Idea of Justice*. Cambridge, MA: Belknap Press of Harvard University Press.

Singer, Peter. 2004. *One World: The Ethics of Globalization*. New Haven, CT: Yale University Press.

Vanderheiden, Steve. 2008a. *Atmospheric Justice*. New York: Oxford University Press.

Vanderheiden, Steve. 2008b. Climate Change, Environmental Rights, and Emission Shares. In *Political Theory and Global Climate Change*, ed. Steve Vanderheiden, 43–66. Cambridge, MA: MIT Press.

Wissenburg, Marcel. 2011. The Lion and the Lamb: Ecological Implications of Martha Nussbaum's Animal Ethics. *Environmental Politics* 20 (3): 391–409.

Young, Iris Marion. 1990. *Justice, Society and Nature: An Exploration of Political Ecology*. London: Routledge.

III

Adjusting Character to a Changing Environment

9

Ethics, Public Policy, and Global Warming

Dale Jamieson

There has been speculation about the possibility of anthropogenic global warming since at least the late nineteenth century (Arrhenius 1896, 1908). At times the prospect of such a warming has been welcomed, for it has been thought that it would increase agricultural productivity and delay the onset of the next ice age (Callendar 1938). Other times, and more recently, the prospect of global warming has been the stuff of "doomsday narratives," as various writers have focused on the possibility of widespread drought, flood, famine, and the economic and political dislocations that might result from a "greenhouse warming"–induced climate change (Flavin 1989).

Although high-level meetings have been convened to discuss the green-house effect since at least 1963 (see Conservation Foundation 1963), the emergence of a rough, international consensus about the likelihood and extent of anthropogenic global warming began with a National Academy Report in 1983 (National Academy of Sciences/National Research Council 1983); meetings in Villach, Austria, and Bellagio, Italy, in 1985 (World Climate Program 1985); and in Toronto, Canada, in 1988 (Conference Statement 1988). In 1988 the Intergovernmental Panel on Climate Change (IPCC) was formed to provide state-of-the-art assessments of climate science. According to the most recent IPCC report (2007), a doubling of atmospheric carbon dioxide from the preindustrial baseline is likely to lead to a 2–4.5 degree centigrade increase in the earth's mean surface temperature. (Interestingly, this estimate is close to that predicted by Arrhenius [1896].) This increase is expected to have a profound impact on climate and therefore on plants, animals, and human activities of all kinds. Moreover, there is no reason to suppose that, without policy interventions, atmospheric carbon dioxide will stabilize at twice preindustrial levels. Moreover, as the human perturbation of natural systems becomes increasingly extreme, the probability of catastrophic climate change increases (National Academy of Sciences, 2002). There are many uncertainties concerning anthropogenic climate change, yet we cannot wait until all the facts are in before we respond. All the facts may never be in. New knowledge may resolve old uncertainties, but it may bring with it new uncertainties. It is an important dimension of this problem that our insults to the biosphere outrun our ability to understand them. We may suffer the worst effects of the greenhouse before we can prove to everyone's satisfaction that they will occur (Jamieson 1991).

The most important point I wish to make, however, is that the problem we face is not a purely scientific problem that can be solved by the accumulation of scientific information. Science has alerted us to a problem, but the problem also concerns our values. It is about how we ought to live, and how humans should relate to each other and to the rest of nature.[1] These are problems of ethics and politics as well as problems of science.

In the first section, I examine what I call the "management" approach to assessing the impacts of, and our responses to, climate change. I argue that this approach cannot succeed, for it does not have the resources to answer the most fundamental questions that we face. In the second section I explain why the problem of anthropogenic global change is to a great extent an ethical problem, and why our conventional value system is not adequate for addressing it. Finally, I draw some conclusions.

Why Management Approaches Must Fail

From the perspective of conventional policy studies, anthropogenic climate change and its attendant consequences are problems to be "managed." Management techniques mainly are drawn from neoclassical economic theory and are directed toward manipulating behavior by controlling economic incentives through taxes, regulations, and subsidies.[2]

In recent years, economic vocabularies and ways of reasoning have dominated the discussion of social issues.[3] Participants in the public dialogue have internalized the neoclassical economic perspective to such an extent that its assumptions and biases have become almost invisible. It is only a mild exaggeration to say that in recent years debates over policies have largely become debates over economics.

The Environmental Protection Agency's draft report "Policy Options for Stabilizing Global Climate" (U.S. Environmental Protection Agency

1989) is a good example. Despite its title, only one of nine chapters is specifically devoted to policy options, and in that chapter only "internalizing the cost of climate change risks" and "regulations and standards" are considered. For many people, questions of regulation are not distinct from questions about internalizing costs. According to one influential view, the role of regulations and standards is precisely to internalize costs, thus (to echo a parody of our forefathers) "creating a more perfect market." For people with this view, political questions about regulation are really disguised economic questions (for discussion, see Sagoff 2004).

It would be both wrong and foolish to deny the importance of economic information. Such information is important when making policy decisions, for some policies or programs that would otherwise appear to be attractive may be economically prohibitive. Or in some cases, there may be alternative policies that would achieve the same ends and also conserve resources.

However, these days it is common for people to make more grandiose claims on behalf of economics. Some economists or their champions believe not only that economics provides important information for making policy decisions but that it provides the most important information. Some even appear to believe that economics provides the only relevant information. According to this view, when faced with a policy decision, what we need to do is to assess the benefits and costs of various alternatives. The alternative that maximizes the benefits less the costs is the one we should prefer. This alternative is "efficient" and choosing it is "rational."

Unfortunately, too often we lose sight of the fact that economic efficiency is only one value, and it may not be the most important one. Consider, for example, the idea of imposing a carbon tax or a market in emissions permissions (i.e., "cap and trade"), as a policy response to the prospect of global warming. What we think of this proposal may depend to some extent on how it affects other concerns that are important to us. Equity is sometimes mentioned as one other such concern, but most of us have very little idea about what equity means or exactly what role it should play in policy considerations.

One reason for the hegemony of economic analysis and prescriptions is that many people have come to think that neoclassical economics provides the only social theory that accurately represents human motivation. According to the neoclassical paradigm, welfare can be defined in terms of preference satisfaction, and preferences are defined in terms of choice behavior. From this, many (illicitly) infer that the perception of self-interest is the only motivator for human beings. This view suggests the following "management technique": If you want people to do something give them a carrot; if you want them to desist, give them a stick.[4]

Many times the claim that people do what they believe is in their interests is understood in such a way as to be circular, therefore unfalsifiable and trivial. We know that something is perceived as being in a person's interests because the person pursues it; and if the person pursues it, then we know that the person must perceive it as being in his or her interests. On the other hand, if we take it as an empirical claim that people always do what they believe is in their interests, it appears to be false. If we look around the world we see people risking or even sacrificing their own interests in attempts to overthrow oppressive governments or to realize ideals to which they are committed. Each year more people die in wars fighting for some perceived collective good than die in criminal attempts to further their own individual interests. It is implausible to suppose that the behavior (much less the motivations) of a revolutionary, a radical environmentalist, or a friend or lover can be revealed by a benefit–cost analysis.

It seems plain that people are motivated by a broad range of concerns, including concern for family and friends, and religious, moral, and political ideals. And it seems just as plain that people sometimes sacrifice their own interests for what they regard to be a greater, sometimes impersonal, good.

People often act in ways that are contrary to what we might predict on narrowly economic grounds, and moreover, they sometimes believe that it would be wrong or inappropriate even to take economic considerations into account. Many people would say that choosing spouses, lovers, friends, or religious or political commitments on economic grounds is simply wrong. People who behave in this way are often seen as manipulative, not to be trusted, without character or virtue. One way of understanding some environmentalists is to see them as wanting us to think about nature in the way that many of us think of friends and lovers—to see nature not as a resource to be exploited but as a partner with whom to share our lives.

Some may think that I have exaggerated the dominance of narrow, economistic approaches to policymaking. Neoclassical economics is in retreat, it might be said, in the wake of the Great Recession of 2008. Paul Krugman has recently reported that some economists are talking about a discipline in crisis (Krugman 2009). Indeed, it might be thought that the hegemony of the neoclassical paradigm has been eroding since the late 1990s, as indicated by Nobel Prizes awarded to Amartya Sen (in 1998), Daniel Kahneman (in 2002), and Elinor Ostrom (in 2009). I am skeptical. The events of 2008 were hardly unprecedented in demonstrating the inability of conventional neoclassical models to predict dramatic changes in the economy. From 1973 to 1974, stocks lost 48 percent of their value; in 1987 the Dow plunged nearly 23 percent in a single day for no apparent reason. A series of "bubbles" that simply could not exist according to some influential theories have occurred throughout the world since the 1980s. Intellectually, there has always been dissatisfaction with the neoclassical paradigm, sometimes even very close to the heart of the discipline.[5] Despite this, the President's Council of Economic Advisors is still in business, unchecked by a Council of Philosophical (or Ethical) Advisors. Public decisions of great consequence continue to be made on the basis of shallow and misleading indicators such as gross domestic product (GDP), rather than on the basis of broader considerations such as quality of life or impacts on fundamental planetary systems.[6] The "straw man" of neoclassical economics seems to me to have quite a lot of blood left in him.

What I have been claiming in this section is that it is not always rational to make decisions solely on narrow economic grounds. Although economic efficiency may be a value, there are other values as well, and in many areas of life, values other than economic efficiency should take precedence. I have also suggested that people's motivational patterns are complex and that exploiting people's perceptions of self-interest may not be the only way to move them. This amounts to a general critique of viewing all social issues as management problems to be solved by the application of received economic techniques. There is a further reason why economic considerations should take a back seat in our thinking about global climate change: There is no way to assess accurately all the possible impacts and to assign economic values to alternative courses of action. A greenhouse warming will have impacts that are so broad, diverse, and uncertain that conventional economic analysis is practically useless.[7]

Consider first uncertainties about the potential impacts, some of which I have already noted. Even if the IPCC report (2007) cited earlier in this chapter is correct in supposing that global mean surface temperatures will increase between 1.1 and 6.4 °C during this century, there is still great uncertainty about the impact of this warming on regional climates. One thing is certain: The impacts will not be homogeneous.

Some areas will become warmer, some will probably become colder, and overall variability is likely to increase. Precipitation patterns will also change, and there is much less confidence in the projections about precipitation than in those about temperature. These uncertainties about regional effects make estimates of the economic consequences of climate change radically uncertain.

There is also another source of uncertainty regarding these estimates. In general, predicting human behavior is difficult. The difficulties are especially acute in the case that we are considering because climate change will affect a wide range of social, economic, and political activities. Changes in these sectors will affect emissions of greenhouse gases, which will in turn affect climate, and around we go again (Jamieson 1988b, 1990). Climate change is itself uncertain, and its human effects are even more radically so. It is for reasons such as these that in general, the area of environment and energy has been full of surprises.

A second reason why the benefits and costs of the impacts of global climate change cannot reliably be assessed concerns the breadth of the impacts. Global climate change will affect all regions of the globe. About many of these regions—those in which most of the world's population lives—we know very little. Some of these regions do not even have monetized economies. It is ludicrous to suppose that we could assess the economic impacts of global climate change when we have such little understanding of the global economy in the first place.

Finally, consider the diversity of the potential impacts. Global climate change will affect agriculture, fishing, forestry, and tourism. It will affect "unmanaged" ecosystems and patterns of urbanization. International trade and relations will be affected. Some nations and sectors may benefit at the expense of others. Moreover, there will be complex interactions between these effects. For this reason we cannot reliably aggregate the effects by evaluating each impact and combining them by simple addition. But since the interactions are so complex, we have no idea what the proper mathematical function would be for aggregating them (if the idea of aggregation even makes sense in this context). It is difficult enough to assess the economic benefits and costs of small-scale, local activities. It is almost unimaginable to suppose that we could aggregate the diverse impacts of global climate change in such a way as to dictate policy responses.[8]

In response to skeptical arguments like the one that I have given, it is sometimes admitted that our present ability to provide reliable economic analyses is limited, but then it is asserted that any analysis is better than none. I think that this is incorrect and that one way to see this is by considering an example.

Imagine a century ago a government doing an economic analysis in order to decide whether to build its national transportation system around the private automobile. No one back then could have imagined the secondary effects: the attendant roads, the loss of life, the effects on wildlife and on communities; the impact on air quality, noise, travel time, and quality of life. Given our inability to reliably predict and evaluate the effects of even small-scale technology (e.g., the artificial heart, see Jamieson 1988a), the idea that we could predict the impact of global climate change reliably enough to permit meaningful economic analysis seems fatuous indeed.

When our ignorance is so extreme, it is a leap of faith to say that some analysis is better than none. A bad analysis can be so wrong that it can lead us to do bad things, outrageous things—things that are much worse than what we would have done had we not tried to assess the costs and benefits at all (this may be the wisdom in the old adage that "a little knowledge can be a dangerous thing").

What I have been arguing is that the idea of managing global climate change is a dangerous conceit. The tools of economic evaluation are not up to the task. However, the most fundamental reason why management approaches are doomed to failure is that the questions they can answer are not the ones that are most important and profound. The problems posed by anthropogenic global climate change are ethical as well as economic and scientific. I will explain this claim in the next section.

Ethics and Global Change

Since the end of World War II, humans have attained a kind of power that is unprecedented in history. While in the past entire peoples could be destroyed, now all people are vulnerable. While once particular human societies had the power to upset the natural processes that made their lives and cultures possible, now people have the power to alter the fundamental global conditions that permitted human life to evolve and that continue to sustain it. While our species dances with the devil, the rest of nature is held hostage. Even if we step back from the precipice, it will be too late for many or even perhaps most of the plant and animal life with which we share the planet (Borza and Jamieson 1990; Thomas et al. 2004, and Bendik-Keymer, chapter 13, this volume). Even if global climate can be stabilized, the future may be one without wild nature

(McKibben 1989). Humans will live in a humanized world with a few domestic plants and animals that can survive or thrive on their relationships with humans.

The questions that such possibilities pose are fundamental questions of morality. They concern how we ought to live, what kinds of societies we want, and how we should relate to nature and other forms of life. Seen from this perspective, it is not surprising that economics cannot tell us everything we want to know about how we should respond to global warming and global change. Economics may be able to tell us how to reach our goals efficiently, but it cannot tell us what our goals should be or even whether we should be concerned to reach them efficiently.

It is a striking fact about modern intellectual life that we often seek to evade the value dimensions of fundamental social questions. Social scientists tend to eschew explicit talk about values, and this is part of the reason why we have so little understanding of how value change occurs in individuals and societies. Policy professionals are also often reluctant to talk about values. Many think that rational reflection on values and value change is impossible, unnecessary, impractical, or dangerous. Others see it as a professional, political, or bureaucratic threat (Amy 1984). Generally, in the political process, value language tends to function as code words for policies and attitudes that cannot be discussed directly.

A system of values, in the sense in which I use this notion, specifies permissions, norms, duties, and obligations; it assigns blame, praise, and responsibility; and it provides an account of what is valuable and what is not. A system of values provides a standard for assessing our behavior and that of others. Perhaps indirectly it also provides a measure of the acceptability of government action and regulation.

Values are more objective than mere preferences (Andrews and Waits 1978; Jamieson 2002, chap. 15). A value has force for a range of people who are similarly situated. A preference may have force only for the individual whose preference it is. Whether or not someone should have a particular value depends on reasons and arguments. We can rationally discuss values, while preferences may be rooted simply in desire, without supporting reasons.

A system of values may govern someone's behavior without these values being fully explicit. They may figure in people's motivations and in their attempts to justify or criticize their own actions or those of others. Yet it may require a theorist or a therapist to make these values explicit.

In this respect, a system of values may be like an iceberg—most of what is important may be submerged and invisible even to the person whose values they are. Because values are often opaque to the person who holds them, there can be inconsistencies and incoherencies in a system of values. Indeed, much debate and dialogue about values involves attempts to resolve inconsistencies and incoherencies in one direction or another.

A system of values is generally a cultural construction rather than an individual one (Weiskel 1990). It makes sense to speak of contemporary American values, or those of eighteenth-century England or tenth-century India. Our individual differences tend to occur around the edges of our value system. The vast areas of agreement often seem invisible because they are presupposed or assumed without argument.[9]

I believe that our dominant value system is inadequate and inappropriate for guiding our thinking about global environmental problems, such as those entailed by climate changes caused by human activity. This value system, as it impinges on the environment, can be thought of as a relatively recent construction, coincident with the rise of capitalism and modem science, and expressed in the writings of such philosophers as Francis Bacon ([1620] 1870), John Locke ([1690] 1952), and Bernard Mandeville ([1714] 1970; see also Hirschman 1977). It evolved in low-population-density and low-technology societies, with seemingly unlimited access to land and other resources. This value system is reflected in attitudes toward population, consumption, technology, and social justice, as well as toward the environment.

The feature of this value system that I will discuss is its conception of responsibility (see also Jamieson 2007a and 2010). Our current value system presupposes that harms and their causes are individual, that they can readily be identified, and that they are local in space and time. It is these aspects of our conception of responsibility on which I want to focus.

Consider an example of the sort of case with which our value system deals best. Jones breaks into Smith's house and steals Smith's television set. Jones's intent is clear: she wants Smith's TV set. Smith suffers a clear harm; he is made worse off by having lost the television set. Jones is responsible for Smith's loss, for she was the cause of the harm and no one else was involved.

What we have in this case is a clear, self-contained story about Smith's loss. We know how to identify the harms and how to assign responsibility. We respond to this breach of our norms by punishing Jones in order to prevent her from doing it again and to deter others from such acts, or we require compensation from Jones so that Smith may be restored to his former position.

It is my contention that this paradigm collapses when we try to apply it to global environmental problems, such as those associated with human-induced global climate change. It is for this reason that we are often left feeling confused about how to think about these problems.

There are three important dimensions along which global environmental problems such as those involved with climate change vary from the paradigm: apparently innocent acts can have devastating consequences, causes and harms may be diffuse, and causes and harms may be remote in space and time.[10]

Consider an example. Some projections suggest that one effect of greenhouse warming may be to shift the Southern Hemisphere cyclone belt to the south. If this occurs, the frequency of cyclones in Sydney, Australia, will increase enormously, resulting in great death and destruction. The causes of this death and destruction will be diffuse. There is no one whom we can identify as the cause of destruction in the way in which we can identify Jones as the cause of Smith's loss. Instead of a single cause, millions of people will have made tiny, almost imperceptible causal contributions—by driving cars, cutting trees, using electricity, and so on. They will have made these contributions in the course of their daily lives performing apparently "innocent" acts, without intending to bring about this harm.[11] Moreover, most of these people will be geographically remote from Sydney. (Many of them will have no idea where Sydney, Australia, is.) Further, some people who are harmed will be remote in time from those who have harmed them. Sydney may suffer in the twenty-first century in part because of people's behavior in the nineteenth and twentieth centuries. Many small people doing small things over a long period of time together will cause unimaginable harms.

Despite the fact that serious, clearly identifiable harms will have occurred because of human agency, conventional morality would have trouble finding anyone to blame. For no one intended the bad outcome or brought it about or even was able to foresee it.

Today, we face the possibility that the global environment may be destroyed; yet no one will be responsible. This is a new problem. It takes a great many people and a high level of consumption and production to change the Earth's climate. It could not have been done in low-density, low-technology societies. Nor could it have been done in societies like ours until recently. London could be polluted by its inhabitants in the eighteenth century, but its reach was limited. Today no part of the planet is safe. Unless we develop new values and conceptions of responsibility, we will have enormous difficulty in motivating people to respond to this problem.[12]

Some may think that discussion about new values is idealistic. Human nature cannot be changed, it is sometimes said. But as anyone who takes anthropology or history seriously knows, our current values are at least in part historically constructed, rooted in the conditions of life in which they developed. What we need are new values that reflect the interconnectedness of life on a dense, high-technology planet.

Others may think that a search for new values is excessively individualistic and that what is needed are collective and institutional solutions. This overlooks the fact that our values permeate our institutions and practices. *Reforming our values is part of constructing new moral, political, and legal concepts.*[13]

One of the most important benefits of viewing global environmental problems as moral problems is that this brings them into the domain of dialogue, discussion, and participation. Rather than being management problems that governments or experts can solve for us, when seen as ethical problems, they become problems for all of us to address, both as political actors and as everyday moral agents.

In this chapter, I cannot hope to say what new values are needed or to provide a recipe for how to bring them about. Values are collectively created rather than individually dictated, and the dominance of economic models has meant that the study of values and value change has been neglected (but see Wolfe 1989; Reich 1988). However, I do have one positive suggestion: *We should focus more on character* and less on calculating probable outcomes.[14] Focusing on outcomes has made us cynical calculators and has institutionalized hypocrisy. We can each reason: Since my contribution is small, outcomes are likely to be determined by the behavior of others. Reasoning in this way, we can each justify driving cars while advocating bicycles or using fireplaces while favoring regulations against them. Even David Brower, the "archdruid" of the environmental movement, owned two cars, four color televisions, two video cameras, three video recorders, and a dozen tape recorders, and he justified this by saying that "it will help him in his work to save the Earth" (San Diego Union, April 1, 1990). More recently, the eleven-day 2009 Copenhagen climate conference produced carbon emissions equal to the annual emissions of 600,000 Ethiopians.

Calculating probable outcomes leads to unraveling the patterns of collective behavior that are needed in order to respond successfully to many of the global environmental problems that we face. When we "economize" our behavior in the way that is required for calculating, we systematically neglect the subtle and indirect effects of our actions, and for this reason we see individual action as inefficacious. For social change to occur, it is important that there be people of integrity and character who act on the basis of principles and ideals.[15]

The content of our principles and ideals is, of course, important. Principles and ideals can be eccentric or even demented. In my opinion, in order to address such problems as global climate change, we need to nurture and give new content to some old virtues such as humility, courage, and moderation and perhaps develop such new virtues as those of simplicity and conservatism. But whatever the best candidates are for twenty-first-century virtues, what is important to recognize is the importance and centrality of the virtues in bringing about value change (Jamieson 2007b; Thompson, chapter 10, this volume; Kawall, chapter 11, this volume; Gardiner, chapter 12, this volume).

The case that I make for the virtues is modest: focusing on character provides a resource that may be useful for solving problems that otherwise seem intractable. Others have made more ambitious claims for the virtues (e.g., Thompson, chapter 10, this volume). They think that acting virtuously is essential to human flourishing, which in turn is understood as resting on a teleological notion of "natural goodness." I am suspicious of such bold metaphysical claims, but I cannot address them here. In any case, their truth is extraneous to the claims that I am making.

Conclusion

Science has alerted us to the impact of humankind on the planet, each other, and all life. This dramatically confronts us with questions about who we are, our relations to nature, and what we are willing to sacrifice for various possible futures. We should confront this as a fundamental challenge to our values and not treat it as if it were simply another technical problem to be managed.

Some who seek quick fixes may find this concern with values frustrating. A moral argument will not change the world overnight. Collective moral change is fundamentally cooperative rather than coercive. No one will fall over, mortally wounded, in the face of an argument. Yet if there is to be meaningful change that makes a difference over the long term, it must be both collective and thoroughgoing. Developing a deeper understanding of who we are, as well as how our best conceptions of ourselves can guide change, is the fundamental issue that we face.

Acknowledgments

This is an updated and revised version of a paper originally published in *Science, Technology and Human Values* 17, no. 2 (Spring 1992): 139–153, based on a lecture given to the American Association for the Advancement of Science in New Orleans, Louisiana, in 1989. I have preserved the central claims, basic argument, and core references because I continue to identify with them. I have also refrained from altering the text to reflect my present voice. As a result, some points are not made as I would make them today. I thank Jeremy Bendik-Keymer and Allen Thompson for guiding the revision, and Michael Glantz who was extremely influential on the original paper. I also thank The Ethics and Values Studies Program of the National Science Foundation (which alas no longer exists, at least in that form) for supporting the research on which the original paper was based.

Notes

1. A claim on which this volume and the humanist conception of adaptation are based.—eds.

2. There are of course other conceptions of management—for example, "adaptive management"—which is used primarily in the domain of natural resource management. See Shockley, chapter 4, this volume, and Hirsch and Norton, chapter 16, this volume, for further discussion.

3. This is as true now as it was in the early 1990s, which we point out in the introduction to this volume when discussing the standard discourse surrounding adaptation.—eds.

4. See Myers 1983 for the view that self-interest is the "soul of modern economic man."

5. See, e.g., Mansbridge 1990, Opp 1989, and Scitovsky 1976. I'm even inclined to enlist Adam Smith as a critic of neoclassical economics.

6. For instance, on the basis of human capabilities or on ecological integrity (see Holland, chapter 7, this volume, and Schlosberg, chapter 8, this volume).—eds.

7. Our inability to perform reliably the economic calculations also counts against the "insurance" view favored by many who favor aggressive action on climate change, but that is another story.

8. Moreover, it is unclear how economic analysis could capture the value of species extinction or radical habitat change in cases where species or habitats are not fungible in terms of exchange value.—eds.

9. Consider the apparent agreement on the environment *not* being a product of our action in Vogel, chapter 15, this volume.—eds.

10. Other important dimensions include nonlinear causation, threshold effects, and the relative unimportance of political boundaries, but I cannot discuss these here, but see Lee 1989 and Jamieson 1991.

11. See Bendik-Keymer, chapter 13, this volume, on "structural wantonness."

12. This call for a new conception of responsibility incites Thompson's argument (chapter 10, this volume) and can be traced through Shockley, chapter 14, this volume; Vogel, chapter 15, this volume; and Hirsch and Norton, chapter 16, this volume, as the collective, institutional, and scalar nature of the problem of responsibility is addressed.—eds.

13. Emphasis added.—eds.

14. Emphasis added.—eds.

15. See Kawall, chapter 11, this volume, on *linking* virtue *to* patterns of collective behavior.

References

Amy, Douglas R. 1984. Why Policy Analysis and Ethics are Incompatible. *Journal of Policy Analysis and Management* 3: 573–591.

Andrews, Richard, and Mary Jo Waits. 1978. *Environmental Values in Public Decisions: A Research Agenda*. Ann Arbor: University of Michigan, School of Natural Resources.

Arrhenius, S. 1896. On the Influence of Carbonic Acid in the Air upon the Temperature of the Ground. *Philosophical Magazine* 41: 237.

Arrhenius, S. 1908. *Worlds in the Making*. New York: Harper & Brothers.

Bacon, F. [1620] 1870. *Works*, ed. James Spedding, Robert Leslie Ellis, and Douglas Devon Heath. London: Longmans Green.

Borza, K., and D. Jamieson. 1990. *Global Change and Biodiversity Loss: Some Impediments to Response*. Boulder: University of Colorado, Center for Space and Geoscience Policy.

Callendar, G. S. 1938. The Artificial Production of Carbon Dioxide and Its Influence on Temperature. *Quarterly Journal of the Royal Meteorological Society* 64: 223–240.

Conference Statement. 1988. The Changing Atmosphere: Implications for Global Security. Toronto, Canada, June 27–30. http://www.cmos.ca/ChangingAtmosphere1988e.pdf (accessed August 24, 2011).

Conservation Foundation. 1963. *Implications of Rising Carbon Dioxide Content of the Atmosphere*. New York: Conservation Foundation.

Flavin, C. 1989. *Slowing Global Warming: A Worldwide Strategy*. Worldwatch Paper 91. Washington, DC: Worldwatch Institute.

Hirschman, Albert. 1977. *The Passions and the Interests*. Princeton, NJ: Princeton University Press.

Intergovernmental Panel on Climate Change (IPCC). 2007. Climate Change 2007: Summary for Policymakers. http://www.ipcc.ch/pdf/assessment-report/ar4/wg1/ar4-wg1-spm.pdf (accessed June 15, 2010).

Jamieson, Dale. 1988a. The Artificial Heart: Reevaluating the Investment. In *Organ Substitution Technology*, ed. D. Mathieu, 277–296. Boulder, CO: Westview.

Jamieson, Dale. 1988b. Grappling for a Glimpse of the Future. In *Societal Responses to Regional Climatic Change: Forecasting Analogy*, ed. Michael H. Glantz, 73–93. Boulder, CO: Westview Press.

Jamieson, Dale. 1990. Managing the Future: Public Policy, Scientific Uncertainty, and Global Warming. In *Upstream/Downstream: New Essays in Environmental Ethics*, ed. D. Scherer, 67–89. Philadelphia, PA: Temple University Press.

Jamieson, Dale. 1991. The Epistemology of Climate Change: Some Morals for Managers. *Society & Natural Resources* 4: 319–329.

Jamieson, Dale. 2002. *Morality's Progress*. Oxford, UK: Oxford University Press.

Jamieson, Dale. 2007a. The Moral and Political Challenges of Climate Change. In *Creating a Climate for Change: Communicating Climate Change and Facilitating Social Change*, ed. S. Moser and L. Dilling, 475–482. New York: Cambridge University Press.

Jamieson, Dale. 2007b. When Utilitarians Should Be Virtue Theorists. *Utilitas* 19 (2): 160–183.

Jamieson, Dale. 2010. Climate Change, Responsibility, and Justice. *Science and Engineering Ethics* 16: 431–445.

Krugman, Paul. 2009. "How Did Economists Get It So Wrong?" New York Times, September 2. http://www.nytimes.com/2009/09/06/magazine/06Economic-t.html (accessed August 24, 2011).

Lee, Keekok. 1989. *Social Philosophy and Ecological Scarcity*. New York: Routledge.

Locke, John. [1690] 1952. *The Second Treatise of Government*. Indianapolis, IN: Bobbs-Merrill.

Mandeville, B. [1714] 1970. *The Fable of the Bees*, trans. P. Harth. Hammersmith, England: Penguin.

Mansbridge, Jane, ed. 1990. *Beyond Self-Interest*. Chicago: University of Chicago Press.

McKibben, W. 1989. *The End of Nature*. New York: Knopf.

Myers, Milton. L. 1983. *The Soul of Modern Economic Man*. Chicago: University of Chicago Press.

National Academy of Sciences/National Research Council. 1983. *Changing Climate.* Washington, DC: National Academy Press.

National Academy of Sciences/National Research Council. 2002. *Abrupt Climate Change, Inevitable Surprises.* Washington, DC: National Academy Press.

Opp, Karl-Dieter. 1989. *The Rationality of Political Protest.* Boulder, CO: Westview.

Reich, Robert, ed. 1988. *The Power of Public Ideas.* Cambridge, MA: Harvard University Press.

Sagoff, Mark. 2004. *Price, Principle, and the Environment.* New York: Cambridge University Press.

Scitovsky, Tibor. 1976. *The Joyless Economy: An Inquiry into Human Satisfaction and Consumer Dissatisfaction.* New York: Oxford University Press.

Thomas, Chris D., et al. 2004. Extinction Risk from Climate Change. *Nature* 427 (January): 145–148.

U.S. Environmental Protection Agency. 1989. Policy Options for Stabilizing Global Climate. Draft report to Congress, ed. D. Lashof and D. A. Tirpak. Washington, DC: GPO.

Weiskel, Timothy. 1990. "Cultural Values and Their Environmental Implications: An Essay on Knowledge, Belief and Global Survival." Paper presented at the American Association for the Advancement of Science, New Orleans, LA.

Wolfe, Alan. 1989. *Whose Keeper? Social Science and Moral Obligation.* Berkeley: University of California Press.

World Climate Program. 1985. *Report of the International Conference on the Assessment of the Role of Carbon Dioxide and of Other Greenhouse Gases in Climate Variations and Associated Impacts.* Report on an international conference held at Villach, Austria, October 9–15. Geneva, Switzerland: World Meteorological Organization.

10

The Virtue of Responsibility for the Global Climate

Allen Thompson

With great power comes great responsibility.
—Stan Lee, creator of *Spiderman*

Act so that the effects of your action are compatible with the permanence of genuine human life.
—Hans Jonas, *The Imperative of Responsibility*

Claims about moral responsibility for climate change are difficult to establish. One problem is that relevant harms and causes are diffuse, another is that actions of individual persons or nations can neither bring about nor prevent the phenomena (Jamieson 1992; revised as chapter 9, this volume). There are also puzzles about what it means for human beings to be responsible for the global climate. Yet thinking clearly about moral responsibility is important not only for negotiating international policy and motivating effective behavior, but also for working out the moral relationship between humanity and the rest of nature (see Hettinger, chapter 1, this volume, and Throop, chapter 2, this volume), especially now that humanity has influence on the geologic scale (Crutzen 2002).

A significant literature on responsibility concerns problems that arise in connection with the global scope of our contemporary society. Sophisticated treatments, developing criteria of agency, freedom, and intentionality in the attempt to make sense of a variety of social and environmental problems, are manifold (e.g., Scheffler 2001; Pettit 2007). My concern, however, is with *responsibility* understood as a good-making trait of character (Williams 2008). Responsibility has long been recognized as an *environmental* virtue but this chapter is about the virtue of being responsible for the (changing) condition of the global climate.

Dale Jamieson argues that thinking clearly about the ethics of global environmental problems like climate change will require a new conception of moral responsibility and he suggests that responding successfully to global environmental challenges will require us to "focus more on character. . . . We need to nurture and give new content to some old virtues," he writes, before suggesting that we may even need to develop some new virtues, suited for the twenty-first century (Jamieson 1992, 148, 150; revised in chapter 9, this volume). I combine these suggestions by arguing that being responsible for the global climate will be a *new* environmental virtue because it is well suited to express human moral goodness in the emerging Anthropocene Epoch (Crutzen 2002).

Global warming threatens our contemporary form of life, the basic ecological conditions to which all life on Earth is adapted, and the moral status of our self-conception as humanity (see Gardiner, chapter 12, this volume). Consider a thought experiment: someday the global environmental crisis will end. Let's be optimistic and imagine that on the other side there will be human beings, living a recognizably civilized form of life. In doing so, we are imagining that human beings could live sustainably as part of the ecological community of life on Earth. By contrast, we know that the dominant "culture of consumerism" is not ecologically sustainable. We may wonder, granting it is possible, just how *different* an ecologically sustainable form of human life would be when compared to the life we know and live today.

Would an ecologically sustainable form of life be different enough to involve novel forms of human goodness? If human activities are resulting in unprecedented anthropogenic changes to the biosphere and global climate system, how might such radical environmental changes and our responsibility for them bear on the environmental virtues? One possibility is that changing environmental conditions will require us to revise the criteria by which relevant virtue or vice terms apply (Kawall, chapter 11, this volume). However, this chapter is about the possibility that novel forms of human goodness regarding the natural environment may emerge, or existing forms may undergo a radical transformation, as we adapt to life in a world where "natural" environments have been significantly transformed by human activities (e.g., the discussion of ecological restoration in part I of this volume). It is about how at least some environmental virtues of the future—the virtues of those living an ecologically sustainable form of life—may be quite different from the environmental virtues of today.

The very idea that environmental and cultural change could result in substantial transformation to the forms of human excellence requires explanation and defense. To this end, Jonathan Lear has provided an account of Plenty Coups (1848–1932), the last principal chief of the Crow Nation, as advancing European settlement devastated traditional Crow forms of life (Lear 2006). I adopt Lear's framework for thinking about the transformation of human virtue through periods of historic change to address issues about responsibility for the global climate. In particular, I focus on the virtue of radical hope, a novel form of courage suited to times of cultural devastation.

First, I review shortcomings that standard conceptions of responsibility suffer in application to global climate change. Then, I summarize Iris Marion Young's conception of political responsibility. Next, I outline Lear's account of Plenty Coups's courage as a radical hope. Finally, I bring the discussion of responsibility and radical hope together, arguing that our thinking about the moral responsibility for global environmental conditions may undergo a radical change, analogous to the transformation for Plenty Coups of courage into radical hope. My view is that a virtue of responsibility, like the virtue of radical hope, is intimately connected with the promise of novel forms of human goodness emerging, in this case from our best response to the global environmental crisis. I describe the general contours of responsibility as a virtue before making suggestions about the shape of one kind in particular: responsibility for the basic conditions of all life on Earth. I argue that such responsibility constitutes a novel human virtue; I compare it with ideas about benevolence and loyalty as virtues of environmental stewardship (Welchman 1999), and defend the view within a neo-Aristotelian framework of natural goodness (Foot 2001).

Standard Responsibility

Under a standard conception of responsibility, guilt or fault is assigned to a person for bringing harm to another. Actions of an agent must bear an important causal relation to a harm that befalls the other person, and if the action that results in harm can also be shown to be voluntary, then it is appropriate to blame the agent for the harmful circumstances. To assign moral responsibility, then, is to find an agent worthy of a particular kind of reaction; in the case of harm, reactions of blame and perhaps punishment. This model derives from legal reasoning designed to assign liability and can apply also to cases in which the agent did not intend to cause the resultant harm, in other words, it is based on a notion of strict liability (Young 2004, 368).

There is a prima facie case that human beings are morally responsible for climate change. In broad terms, global climate change is caused by an increase in atmospheric concentrations of carbon dioxide and other greenhouse gases, which in turn are caused primarily by human beings engaged in many and various activities. More, significant present *harms* can be attributed to the changes in global climate and significantly greater harms will befall future generations, unless anthropogenic atmospheric forcings are significantly curtailed (IPCC 2007).

The case for moral culpability depends on showing that human beings undertake the relevant acts *voluntarily.* At the level of individual actions, those typically undertaken by relatively well-off persons in developed nations (such as setting the thermostat to a certain range), we have no reason to doubt that people act voluntarily, exercising choice over options. This is less clear when we consider people living at or near subsistence conditions—individuals who may not have real options regarding the carbon cost of their daily activities (Shue 1993).

It's more typical to speak about humanity *as a whole* as responsible for climate change and parse the question of acting voluntarily in terms of knowledge that doing so causes global climate change. After the 1990 Intergovernmental Panel on Climate Change (IPCC) report, it was no longer plausible to deny that we understood (roughly) the climate consequences of the status quo. Considered alone, knowledge that actions cause harm is not sufficient to establish the actions are done voluntarily. But knowledge is a necessary condition and, wed to an optimistic view of human autonomy, it's plausible to believe that humanity was *not* morally responsible *only* so long as we remained ignorant.

When, as in this case, harms connected with the status quo are significant, the harms associated with acting otherwise would have to be comparable for such an appeal to exculpate. In fact they are not (Stern 2007). Yet for almost twenty years, humanity has failed to take serious action to mitigate climate change, action we could have taken once we knew about the risks. In this sense, it is not uncommon to believe that human beings are morally responsible for global climate change.

It is important to be clear about responsibility because of its place in justifying blame, punishment, retribution, or moral obligations. Even if we accept that human beings are morally responsible for climate change, significant difficulties arise in the effort to make more precise sense of this claim. On Jamieson's analysis, the standard conception of responsibility presupposes that "harms and their causes are individual, that they can be readily identified, and that they are local in space and time" (Jamieson 1992, 148). But with many global environmental problems, including climate change, harms and causes are diffuse; they are not easily identified, and they are remote across space and time. The contributory acts of individuals are undertaken without malice and indeed are typically quite harmless in isolation.[1]

One central reason that many global environmental problems are not easily articulated in terms of individual moral responsibility is that they have the structure of collective-action problems. Characteristic of such, each individual has incentive to behave in a way that if most others also behave in that way, the consequences will be suboptimal for each according to her own preferences. Given that the harmful consequences are brought about only by the aggregate of what most do, it does not make sense to hold an *individual* responsible for the outcome. As this will be true for every individual in the group, "today we face the possibility that the global environment may be destroyed, yet *no one* will be responsible" (Jamieson 1992, 149, revised as chapter 9, this volume, italics added).

If the cause of climate change is the aggregate actions of many, perhaps the appropriate sense of moral responsibility is a collective one. In one sense, collective responsibility is only the sum of the individual responsibility of the members of a group (Feinberg [1968] 1970, 683). Despite the impotence of anyone alone to alter the global climate, given the current energy regime, each individual does make *some* contribution to global greenhouse gas emissions, and we typically assign *some* degree of individual moral responsibility for the informed and voluntary lifestyle choices of those living well above subsistence levels. In this sense, collective human responsibility for global climate change is an aggregate notion, distributed proportionately to those individuals whose personal greenhouse gas emissions exceed a fair share of the planet's capacities to process them without altering global climate stability (Shue 2000).

One problem is that negotiating international policy seems to require a sense of collective responsibility that is rightly attributable to a unified whole.[2] Were the United States to admit responsibility for its contribution to the accumulated greenhouse gases, this would not be shorthand for the proportionate, personal responsibilities of its individual citizens (many of whom are dead). As citizens of one nation bound together by shared democratic institutions and a shared cultural narrative, at some level we are equally responsible for the policies and practices that led to these consequences. Were it otherwise, the collective sense in which the

United States bears moral responsibility would stand in direct conflict with values of individual responsibility and justice. How could it be just that a whole nation be sanctioned when the relevant harms were caused more by some than others?

A second problem with this distributed sense of collective responsibility is that it's not obvious any individual thereupon has an obligation to reduce her personal energy consumption. Let's grant that if a person's actions cause harm, that person has reason not to act. But an individual's actions (even over a lifetime) are neither necessary nor sufficient to cause climate change. While she may be individually responsible for her consumer behaviors, her actions have not caused the harm and so she has no reason to reduce her personal emissions (Sinnott-Armstrong 2005). So, this collective but distributed sense of group fault leaves individuals with no obligation to mitigate their personal emissions, and it fails to assign responsibility to the group as a whole when serious harms are genuinely the product of many hands.[3]

Another sense of collective responsibility is nondistributive: responsibility is attributed to the group *as a whole,* distinct from its members (Feinberg [1968] 1970, 687). On this view, the claim that human beings are responsible literally attributes responsibility to humanity as a whole. Thus, it is consistent with the view that no individuals are to blame. But there are problems here, too. First, if people are not individually responsible for the harms of global warming, then it's difficult to see how they could be blameworthy for failures to limit personal emissions. However, there are environmentalist intuitions that individuals can be morally blameworthy for making unnecessary or excessive contributions to the build-up of global greenhouse gases, even if they have no obligation to do otherwise. There may also be good consequentialist reasons to blame them (Jamieson 2007).

Second, a nondistributive conception of group fault is itself controversial on several grounds. On one hand, *methodological* individualists object that collectives cannot meet conditions of agency, such as forming intentions, required to make the ascriptions of voluntary action and blameworthiness intelligible (Narveson 2002). While these kind of metaphysical criticisms may be blunted in cases of highly organized groups exemplified by corporations or some nation-states, it's difficult to see how the collection of *all human beings* could plausibly be considered an agent in the relevant sense. On the other hand, *normative* individualists object that assigning collective responsibility has unjust practical implications when individuals who in fact have not contributed, or contributed less, to the harm are included on equal terms under group sanction or blame (French 1984; May 1987; see also Smiley 2005). It is well known that citizens of the developed, industrial nations make far and away the greatest per capita contributions to global greenhouse gas concentrations. It seems unfair that impoverished individuals living in underdeveloped nations should be found equally culpable, solely on the basis of being equally part of humanity.

Let's review. It's plausible that human beings, in some sense, meet the conditions of responsibility for global climate change. But there is not an obvious sense in which either individuals or collectives can be understood to bear the moral responsibility. Yet international policy negotiations offer a pragmatic need for some sense of *collective* responsibility and there is a need to understand the status of *individual* responsibility in order to think clearly about personal obligations and help motivate effective action. I will not address further the issue of national, collective responsibility. Instead I focus on individual responsibility for the harms caused by what many, many human beings do. Further, I try to make a case that understanding responsibility for global climate *as a virtue* can help us think about new forms of human moral goodness.

Political Responsibility

A conception suitable to the problem would have to make sense of how humans are both individually and collectively morally responsible for global climate change. Thus, a core issue concerns the relationship between individuals and the group. A political analysis would examine the relation of a citizen to her state. Framed to deal with issues of international social justice, Iris Marion Young distinguishes a model of *political* responsibility from the standard *liability* model, because people "have difficulty reasoning about individual responsibility with relation to outcomes produced by large scale social structures in which millions participate, but of which none are the sole or primary cause" (Young 2004, 374). This is an apt description of our problem.

Young acknowledges Hannah Arendt as providing her point of departure. Arendt identifies political responsibility as a form of nondistributive collective responsibility, the assumption of which does not turn on individual fault but derives from "my membership in a group (a collective) which no voluntary act of mine can dissolve" (Young 2004, 375). For Arendt, the paradigm group is the polity, such as a republic or nation-state. Young, however, argues that the social and economic relations that underwrite the political responsibility citizens owe to one another have, with globalization, broadened beyond national boundaries (377). Young characterizes political responsibility as follows:

(1) Unlike a blame model of responsibility, political responsibility does not seek to mark out and isolate those to be held responsible, thereby distinguishing them from others, who by implication then are not responsible. . . . Most accounts of collective responsibility aim to distinguish those who have done harm from those who have not. . . . Political responsibility, on the other hand, is a responsibility for what we have *not done.*

(2) In a liability conception of responsibility, what counts as a wrong for which a perpetrator is sought . . . is generally conceived as a deviation from a baseline. Implicitly we assume a normal background situation that is morally acceptable, if not ideal. . . . A concept of political responsibility. . ., on the other hand, evaluates not a harm that deviates from the normal and acceptable, but rather often brings into question precisely the background conditions that ascriptions of blame or fault assume as normal.

(3) Political responsibility . . . differs from a liability model of responsibility in being more forward-looking than backward-looking. . . . [It] seeks not to reckon debts, but aims rather to bring about results, and thus depends on the actions of everyone who is in a position to contribute to the results.

(4) Political responsibility is relatively open with regard to the actions that count as taking up the responsibility. . . . Like duties, responsibilities carry a burden and an obligation; carrying out responsibilities is not a matter of mere beneficence. Unlike duties, however, responsibilities carry a considerable discretion; one *must* carry out one's responsibilities, but *how* one does so is a matter for judgment. . . . One has fulfilled the duty if one has preformed the required actions. Carrying out a responsibility, on the other hand, consists in seeking to bring about a specified outcome.

(5) Political responsibility . . . is a shared responsibility . . . [the concept of which is thus] distinct from the concept of collective responsibility in that the former is a distributed responsibility whereas the latter is not. . . . [It is] a personal responsibility for outcomes, or the risk of harmful outcomes, produced by a group of persons. Each is personally responsible for the outcome in a partial way, since he or she alone does not produce the outcomes; the specific part that each plays in producing the outcome cannot be isolated and identified, however, and thus the responsibility is essentially shared. (Young 2004, 377–380)

At least in outline, Young's conception of political responsibility is well suited for modeling human moral responsibility for global climate change. Finding some to blame, it does not thereby exculpate others; people are guilty primarily for what they have *not* done; wrongs are not *contrasted* with the status quo; rather, the moral status of background conditions (e.g., consumerism) is brought into question; every and all contributors are accountable for an *outcome*; political responsibility does not seek retribution for past actions; it generates obligations that cannot be satisfied simply by acting as required and which may vary from case to case. Perhaps most important, political responsibility is a *shared* responsibility, distributed to each because the contributions of each member "cannot be isolated and identified." Responsibility for the outcome does not belong strictly to some individuals or to some collectives: humans have a shared moral responsibility for global climate change.

The idea that human responsibility for climate change should be modeled on the social conception of a shared, political responsibility appears also in Vogel's analysis of alienation from the environment and the essentially social structure of the problem (chapter 15, this volume). However, if the key is found in an account of *shared* responsibility built on an intimate relationship between individuals *as* a group, we should note that political bonds do not exhaust the type. One alternative is to consider the relation of an individual to its natural kind. From the perspective of an Aristotelian ethical naturalism the relation of a human being to its natural kind is articulated in terms of virtue and natural goodness (Foot 2001). Each of us participates in our common humanity in a way no one can voluntarily disavow. Thus, in the discussion that follows I consider moral responsibility modeled on Young's political conception as a moral virtue.

Looking beyond the model of political responsibility is motivated also by the sheer magnitude of anthropogenic climate change as a moral problem. It is not hyperbole to say we stand at the beginning of an era the likes of which humanity has never known.[4] The impending changes to the conditions under which all life on Earth exists threaten to be radical and utterly unprecedented, perhaps requiring equally radical and adaptive changes to our system of values, even to our understanding of humanity and moral assessment of who we are (see Jamieson, chapter 9, this volume, and Gardiner, chapter 12, this volume). This is not merely an economic or a political problem. What could it mean to anticipate or undergo such radical moral change? One answer is offered in Lear's account of how the courage of Plenty Coups and the Crow was transformed through a time of cultural devastation.

Radical Hope

An analogy with the Crow begins with what many already accept: the dominant consumer culture is not ecologically sustainable. By "consumerism," I mean a set of attitudes and values leading people always to high levels of consumption and orienting them to find meaning and satisfaction in life largely through the practices of purchasing new consumer goods (Goodwin et al. 2007). Modern practices of consumption are depleting natural resources faster than they can regenerate and research supports the claim that we have already overshot the global carrying capacity (Rockstrom et al. 2009). Although some of us are more consumptive than others, I presume that almost everyone in the developed world exhibits features of our cultural norm. By "sustainable," I only mean if a practice is not sustainable, then continuing with it now effectively prohibits others from doing so in the future. Understood this way, it follows that if consumer culture is not sustainable, we can predict its demise.

Like the Crow, we may soon face significant cultural change.[5] How could the end of consumer culture affect our understanding of the environmental virtues? Cultural relativism is a meta-ethical view that the truth of moral claims, including claims about the virtues, is dependent upon contingent cultural features. On this view, significant cultural change—such as the end of consumerism—can directly affect what character traits count as genuine virtues. But would allowing that deep cultural change can significantly alter the moral virtues commit us to some form of cultural relativism?

The answer is no. We can recognize a significant role for culture in shaping how the virtues are understood in a particular time and place, yet remain nonrelative about true moral judgments. To do so, we distinguish between thick and thin conceptions of virtue. The specific way that members of a particular culture exhibit a trait of character and the cultural practices and frameworks in which the trait is understood as exemplifying excellence and the good life represent a culturally thick conception of virtue. By stripping away culturally dependent frameworks and specific practices, we begin to thin out the characterization of a virtue, getting to a description that transcends culturally bound specification and applies to human beings more universally. It's not difficult to recognize this distinction; difficulty lies in recognizing how to act virtuously without one's culturally thick conception of living well.

The Crow were nomadic hunters, a warrior culture developed over hundreds of years to a life of fierce competition with other tribes for control of a shifting territory and access to buffalo. In traditional Crow life, "*everything* counted as either hunting or fighting or as preparing to hunt and fight" (Lear 2006, 40). Unsurprisingly, the Crow revered and celebrated the virtue of courage. Their paradigm of courage was the practice of planting a coup-stick, the primary use of which was to mark a boundary, a place past which no non-Crow shall be tolerated (23). As a youth, Plenty Coups exhibited great courage in battle against the Sioux, but he became chief when Crow survival was threatened more directly by European settlement. Eventually, between 1882 and 1884, the Crow were moved onto a reservation, where intertribal warfare and all forms of counting coups were impossible. Here is the puzzle: when the Crow paradigm of courage was practically impossible, how could Plenty Coups lead courageously?

According to Lear, Plenty Coups's leadership was a courageous response to the radical changes befalling the Crow. Under the new conditions, still emerging and unknown, leading his people to appropriate adaptation would require an outstanding exercise of practical wisdom. But acting wisely in the midst of utter cultural devastation required Plenty Coups to anticipate a future for the Crow that he did not yet know how to think about. On Lear's interpretation, the Crow did not have the conceptual tools to grasp the changes that were underway nor, more important, how they could go on *as the Crow.*

In response, Lear suggests that there might be "a certain plasticity deeply embedded in a culture's thick conception of courage" and defends the view that there may be "ways in which a person brought up in a culture's traditional understanding of courage might draw upon his own inner resources to broaden his understanding of what courage might be. In such a case, one would begin with a culture's thick understanding of courage; but one would find ways to *thin it out*: find ways to face circumstances courageously that the older thick conception never envisioned" (Lear 2006, 65, italics added).

Plenty Coups was committed to a goodness that transcended his understanding, and was hopeful his people would be able to "get the good back"—not only would they survive the destruction of their traditional forms of life, but they would return again to flourish in the presently unimaginable new world. Plenty Coups thus embodied a virtue of radical hope, "basically the hope for revival: for coming back to life in a form that is not yet intelligible," the virtue of seeing that the world's goodness outstrips the ability of one's culture to capture it. Radical hope is against despair, even in the face of a well-justified despair. It is the idea that an inadequate grasp of the good should not lead one to believing it is not to be hoped for. "At a time of radical historical change," Lear writes, "the concept of courage will itself require new forms" (Lear 2006, 118).

[handwritten margin note: Seems like it is the Crow type of courage, "the transcendent"]

A product of imaginative excellence and practical wisdom, radical hope allows courage to be manifest in situations where one has an outdated conception of living well. Traditionally, courage concerns the willingness to risk significant harm defending some worthy good. However, radical hope is a form of courage at the end of goodness, a steadfastness underpinning action on the idea that someday the good will return in a presently unimaginable form. Radical hope is thus a distinctly novel form of courage, exhibiting commitment to some unknown but worthy conception of the good life, to an unknown form of flourishing. That Plenty Coups's radical hope was not merely wishful thinking depends on the possibility that a new form of flourishing for the Crow *would* emerge. Culturally thick conceptions of what it was to live well *as a Crow* were dependent on the continued intelligibility of their traditional forms of life, which were now impossible.

From an Aristotelian perspective, any genuine virtue is a substantive manifestation of the good. Thus, genuinely new forms of human goodness entail new virtues, just as genuinely new virtues implicate new forms of the good life. Radical hope is one special case; as a new form of courage, radical hope is conceptually tied to the *promise* of a new form of the good life, a new form of flourishing *as the Crow* that will arise to fill the vacuum created by the destruction of their traditional form of life.

I have argued in prior work that in light of ubiquitous human influence on the global environment, radical hope *for an environmentalist* requires commitment to the emergence of a new but noninstrumental form of goodness in nature, goodness distinct from traditional conceptions of value in nature as autonomy from human influence. To deny that nature's noninstrumental goodness must be connected with its autonomy, I claimed, is to affirm the possibility that a new conception of a *natural* goodness—the object of an environmentalist's radical hope—is connected with human responsibility for the natural world (Thompson 2010).[6]

Here, I use the model of radical hope to focus on the virtue of responsibility and the emergence of a new form of *human* goodness. Roughly, my view is that what we appeal to in explaining global climate change as anthropogenic, meaning the human capacity to affect the global environment, gives rise to the possibility of a new form of human goodness. I take it for granted that, as a matter of fact, humanity now has the role of managing the global biosphere. We were neither designed nor destined for this; only the contingent course of history has made it so. The claim

[handwritten note at bottom: Just happened]

is descriptive, not normative; it's not that we *ought* to fill this role, but simply that we do. Human beings are now managers of the planet in the sense that collectively our actions determine the basic conditions for the existence of all life on Earth. The virtue of prudence, it seems obvious, would suggest we understand this role as one of stewardship. This is the moral significance of what it means to say that we live in the Anthropocene Epoch.

If we accept that humanity now has the role of managing planetary stewardship, then we can recognize the demands of this role can be met well or badly. Thus, we see the possibility of a new form of human goodness. Satisfying well the demands of this role will require human beings to develop suitable and correspondingly new traits of character among which, I argue, is a special *virtue of responsibility*.

Responsibility as a Virtue

Standard conceptions of responsibility tend to be backward looking, focusing on questions of freewill or rational agency and the criteria for assigning guilt and blame. Young's conception of political responsibility, rooted in the interrelations of world citizenship, is responsibility *shared* by members of a collective. It is outcome oriented, holding each member accountable for her partial contribution to a consequence produced only by the whole. One's status as a member of this group cannot be annulled by an act of will, and the nature of this bond contributes to the sense in which responsibility is shared.

Then there is a sense in which we praise some people as more responsible than others, despite being equally free or accountable. Garrath Williams offers an account of responsibility as a character virtue and cites the secular, political contexts of mutual accountability out of which use of the term "responsibility" historically arose. He then goes on to say that "we most often use the term without particular references to citizenship and its duties" (Williams 2008, 458). This broad use supports the view that responsibility is also a *human* virtue, not just a *civic* virtue.

What do we mean when we praise someone as responsible? A responsible person, Williams notes, is reliable, exhibiting commitment over time; she has initiative and judgment, she can be trusted and exercises discretion. How might we define the virtue? Williams refers to Weber: "a person . . . 'must bear the (foreseeable) *consequences of his actions*,' which requires that he be able to face realities 'with inner composure and calm.'" He also refers to Fingarette: "responsibility 'emerges where the individual accepts as a matter of personal concern something which society offers to his concern.'" Finally, he claims that "responsibility represents the readiness to respond to a plurality of normative demands" (Williams 2008, 459).

The virtue of responsibility also connects with a standard notion of responsibility that seeks to assign guilt, fault, and blame. "With some circularity," Williams writes, "one might say that [the virtue of] responsibility suggests an agent who lives up to her . . . position within a division of responsibilities and within relations of mutual accountability" (Williams 2008, 459). Recognizing distinct spheres is to acknowledge a conception of responsibility as the expectations or obligations associated with a specific *role*. Thus, we can understand the virtue of responsibility as a capacity with respect to taking care of multiple requirements that accompany one's role, to be willing and able to do one's best (with a reasonable chance of success) to ensure that things come out right.

If we accept William's generic definition of responsibility and understand the model of political responsibility outlining the "plurality of normative demands" that responsible citizens manage well, we have the outline of a *virtue* of political responsibility as a settled disposition to respond and manage well the satisfaction of various obligations derived from one's political responsibility. Insofar as a person plays the role of global citizen, she is made good by the virtue of political responsibility. My purpose is to suggest we consider the virtue of responsibility in terms of humanity's role as planetary stewards. If humanity actually has this role, then good humans will be made good by the relevant virtues of stewardship, including an *environmental* responsibility, which disposes them to meet well the plurality of normative demands confronting one and all who share in being accountable for the global climate, the basic conditions supporting life on Earth.

In the secular tradition, the claim that humanity has the role of planetary stewardship is now backed by the best of our atmospheric sciences. Because every human being shares in our common humanity, environmental responsibility is essentially a *shared* moral responsibility. Humans are made good by the virtue of environmental responsibility. According to an Aristotelian naturalism, individuals are evaluated with reference to standards of human goodness. A part of human goodness in the Anthropocene, I argue, is the disposition to meet well the plurality of normative demands derived from humanity's role responsibility qua managers of the Earth's global climate and basic ecological conditions.

Recognizing the good of humanity to include responsibility for the global climate and outlining it on a model of shared political responsibility leaves unaddressed important questions about the distribution of this responsibility to individuals. This is an important lacuna, because it is of consequence to understanding what environmental responsibility demands of a particular person. Clearly, no one individual alone is or could be accountable for the stewardship of life on Earth. Yet neither is anyone off the hook. The responsibility is essentially shared; so the corresponding excellence of character presents a standard against which anyone, as a member of humanity, can be morally assessed. We will want to know what the virtue of environmental responsibility (including the global climate) demands of each person.

We cannot answer this by appeal to principle. Theoretic under-determination here is implied by three features of my view: Aristotelian ethics, political responsibility, and radical hope. Aristotelian ethics generally leaves the demands any virtue makes of an individual dependent upon particulars of the agent and her specific circumstances; no less should be expected here.[7] According to Young, shared political "responsibilities carry a considerable discretion; one *must* carry out one's responsibilities, but *how* one does so is a matter for judgment" (Young 2004); the same will carry over to environmental responsibility. Finally, details of what human goodness qua planetary steward amounts to elude us for now; like the Crow, we lack a conceptual grasp of the emerging form of our flourishing. Today we need radical hope, commitment to the idea that responsibility for life on Earth constitutes a part of human flourishing, although we don't yet understand precisely how to flourish in that way.

but what more?

Conclusion

It's not uncommon to claim that humans ought to be good stewards of the natural environment, or that virtues have an important place in performing this role well. Jennifer Welchman, for example, defends benevolence and loyalty among the virtues of stewardship (Welchman 1999). I understand "virtues of stewardship" as a collection of character traits that are excellences of character for individuals who have the role of environmental steward, which I have claimed is a responsibility shared by all humanity. So, the virtue of environmental responsibility is among the virtues of stewardship, yet is among them unique. *Environmental responsibility is the quality of managing well the satisfaction of a plurality of normative demands derived from our accountability for the basic conditions of life on Earth.* Hence, environmental responsibility is a disposition to address well the demands on one's character that the role of stewardship entails; it is a character excellence regarding the possession of other specific virtues of stewardship, such as benevolence and loyalty. The environmentally responsible person will be disposed to acquire and exercise the particular virtues of environmental stewardship.

On *eudaimonistic* views, a virtue is a character trait that is constitutive of human flourishing. I understand judgments of human flourishing according to a *natural goodness approach*, whereby evaluative judgments of human character and rationality depend "directly on the relation of an individual to the 'life form' of its species" (Foot 2001, 26).[8] Environmental virtues, then, are excellences regarding various relations to the organisms and ecosystems of the terrestrial biosphere where our specifically human lives unfold. Environmentally good human beings recognize what is fine in and about the natural world and are habitually disposed to act well regarding these values, thus enabling the goodness of nature to have a substantive role in human flourishing, just as the goodness of friends and human community—via the relevant interpersonal virtues—can have such a role.

The status of environmental responsibility (for the global climate) as a human character virtue depends on the human form of life now including the role of planetary steward. Often, Aristotelian ethical naturalism is mistaken for what McDowell calls "bald naturalism," a mistaken view attempting to somehow defend the rational justification of ethical norms as relying on the objectivity of scientific biology. The patterns and forms of life characteristic of human "nature" are thought to be those derived historically, consistent with or even informed by our evolutionary heritage (McDowell 1995). The view I am advancing, if at all plausible, shows the natural goodness approach need not be so constricted. Indeed, human life has been informed by our biological evolution, of course, but takes a form delivered in part by the contingencies of history and culture. The power humanity has today to alter the global biosphere is an objective, natural fact that we cannot simply disavow. The flourishing of life on Earth, including human life, depends on our exercising this power well, thus to flourish qua human we need the character traits that enable us to manage this role well. I have argued that environmental responsibility is key among them, a form of natural goodness morally befitting our humanity.

Acknowledgments

I want to thank Dale Jamieson and Garrath Williams for helpful feedback and Jeremy Bendik-Keymer for directing me toward Iris Marion Young's work on political responsibility.

Notes

1. Compare Bendik-Keymer, chapter 13, this volume.

2. For an alternative, cosmopolitan conception of climate ethics that put an emphasis on individual rather than national greenhouse gas emissions, see Harris 2010.

3. John Nolt (2011) defends the view that through their personal greenhouse gas emissions individuals are causally responsible for significant harm to future human beings.

4. See Higgs, chapter 4, this volume, on limits of our "ecological imagination."

5. This section draws substantively on Thompson 2010.

6. Compare Hettinger, chapter 1, this volume, on this point.

7. I am sympathetic with particularist views, such as Dancy 2004.

8. I outline and defend a version of the natural goodness approach in Thompson 2007. See also Sandler 2007.

References

Crutzen, P. J. 2002. Geology of Mankind. *Nature* 415, 23. http://www.nature.com/nature/journal/v415/n6867/full/415023a.html (accessed March 9, 2009).

Dancy, Jonathan. 2004. *Ethics without Principles*. Oxford, UK: Clarendon Press.

Feinberg, Joel. [1968] 1970. Collective Responsibility. In *Doing and Deserving: Essays in the Theory of Responsibility*, 222–251. Princeton, NJ: Princeton University Press.

Foot, Philippa. 2001. *Natural Goodness*. Oxford, UK: Clarendon Press.

French, Peter. 1984. *Collective and Corporate Responsibility*. New York: Columbia University Press.

Goodwin, N. R., J. A. Nelson, F. Ackerman, and T. Weisskopf. 2007. Consumer Society. In *Encyclopedia of Earth*, ed. Cutler J. Cleveland. Washington, DC: Environmental Information Coalition, National Council for Science and the Environment. http://www.eoearth.org/article/Consumer_society (accessed August 10, 2008).

Harris, Paul. 2010. Misplaced Ethics of Climate Change: Political vs. Environmental Geography. *Ethics, Place, and Environment* 13 (2): 215–222.

Intergovernmental Panel on Climate Change (IPCC). 2007. *Fourth Assessment Report (AR4): Mitigation of Climate Change*. Cambridge, UK: Cambridge University Press.

Jamieson, Dale. 1992. Ethics, Public Policy, and Global Warming. *Science, Technology & Human Values* 17 (2): 139–153.

Jamieson, Dale. 2007. When Utilitarians Should Be Virtue Theorists. *Utilitas* 19 (2): 160–183.

Jonas, Hans. 1985. *The Imperative of Responsibility: In Search of an Ethics for the Technological Age*. Chicago: University of Chicago Press.

Lear, Jonathan. 2006. *Radical Hope: Ethics in the Face of Cultural Devastation*. Cambridge, MA: Harvard University Press.

May, Larry. 1987. *The Morality of Groups*. Notre Dame, IN: Notre Dame Press.

McDowell, John. 1995. Two Sorts of Naturalism. In *Virtues and Reasons*, ed. R. Hursthouse, G. Lawrence, and W. Quinn, 149–180. Oxford: Clarendon Press.

Narveson, Jan. 2002. Collective Responsibility. *Journal of Ethics* 6: 179–198.

Nolt, John. 2011. How Harmful Are the Average American's Greenhouse Gas Emissions? *Ethics, Policy & Environment* 14 (1): 3–10.

Pettit, Philip. 2007. Responsibility Incorporated. *Ethics* 117: 171–201.

Rockstrom, John, et al. 2009. A Safe Operating Space for Humanity. *Nature* 461 (24): 472–475.

Sandler, Ronald. 2007. *Character and Environment*. New York: Columbia University Press.

Scheffler, Samuel. 2001. Individual Responsibility in a Global Age. In *Boundaries and Allegiances: Problems of Responsibility and Justice in Liberal Thought*, 32–47. Oxford, UK: Oxford University Press.

Shue, H. 1993. Subsistence Emissions and Luxury Emissions. *Law & Policy* 15: 39–59.

Shue, H. 2000. Climate. In *A Companion to Environmental Philosophy*, ed. Dale Jamieson, 449–459. Malden, MA: Blackwell Publishing.

Sinnott-Armstrong, W. 2005. "It's Not My Fault": Global Warming and Individual Moral Obligations. In *Perspectives on Climate Change: Science, Economics, Politics, Ethics*, ed. W. Sinnott-Armstrong and R. B. Howarth, 221–253. Amsterdam: Elsevier.

Smiley, Marion. 2005. "Collective Responsibility." *Stanford Encyclopedia of Philosophy*. http://plato.stanford.edu/entries/collective-responsibility/ (accessed October 10, 2009).

Stern, Nicholas. 2007. *The Economics of Climate Change: The Stern Review*. Cambridge, UK: Cambridge University Press.

Thompson, Allen. 2007. Reconciling Themes in Neo-Aristotelian Metaethics. *Journal of Value Inquiry* 41 (2): 245–264.

Thompson, Allen. 2010. Radical Hope for Living Well in a Warmer World. *Journal of Agricultural & Environmental Ethics* 23 (1): 43–59.

Welchman, Jennifer. 1999. The Virtues of Stewardship. *Environmental Ethics* 21: 411–423.

Williams, Garrath. 2008. Responsibility as a Virtue. *Ethical Theory and Moral Practice* 11 (4): 455–470.

Young, Iris Marion. 2004. Responsibility and Global Labor Justice. *Journal of Political Philosophy* 12 (4): 365–388.

11
Rethinking Greed

Jason Kawall

Greed is often thought to be a particularly common and troubling vice in contemporary, market-driven societies.[1] The negative effects of greed seem wide-ranging and severe: environmental harms afflicting current generations of humans (and nonhumans), exploitation of workers across the world, weakened communities, a turning away of the greedy themselves from genuine self-improvement and well-being, and, perhaps most important for present purposes, potentially devastating impacts upon future generations. With climate change—largely driven by growing human consumption—future generations face drastically changing ecosystems, massive species loss, radically changing regional climates, extreme weather systems, coastal flooding and erosion, and still further adverse impacts. Moral burdens are being placed on future generations that will need to make difficult, possibly tragic decisions in the face of a radically changed world.

Those wealthy by global standards, in particular, may seem to exhibit greed in their ongoing demand for a wide range of goods. They seem to be key drivers of the consumption that is fuelling potentially devastating environmental changes. Yet we might wonder whether the majority of the globally wealthy are, in fact, greedy; after all, the common image of the greedy person is of one who is obsessed with getting more material goods, or of the miser counting his coins. Are we really like this?

In what follows I attempt to clarify the nature of the vice of greed, focusing on what can be called "modest greed." Agents who are modestly greedy do not long for material goods or wealth with intense desires. Rather, they have quite modest desires, but ones whose satisfaction they pursue excessively relative to other goods. Modest greed will emerge as a particularly troubling vice for those facing a changing world—one that will require us to reassess our actions, our beliefs, and our understanding of the virtues.

Defining Greed

Greed is a vice of *disproportionality*: greedy agents pursue objects to a degree that is *disproportionate* to their value relative to other goods that the agents could be pursuing, or to the harms associated with their pursuit.[2] For example, there is pleasure and value in enjoying a doughnut, but the greedy person desires doughnuts disproportionately to other valuable goods, such as his health. Similarly, a pursuit of wealth in itself need not be greedy, for wealth has positive instrumental value. But when an agent pursues wealth to the detriment of the goods of friendship, the development of other interests and talents, or even moral constraints, her pursuit becomes greedy.

There can be greed for most any good. Thus, we could speak of being greedy for praise or for knowledge insofar as we can pursue such goods excessively relative to other goods. But our discussion will focus on paradigmatic cases of greed, those with material goods or wealth as their objects.

Consider, then, the following account of greed with respect to a given good. Intuitively, the account can be seen as divided into two parts: clauses (1) through (3) capture the excessive pursuit of objects definitive of greed, while clauses (4) and (5) capture greed's broader manifestations in our attitudes and actions:

One is greedy with respect to a good or set of goods to the extent that one (1) excessively pursues this good or set of goods, or would do so in relevant conditions, due to either (2a) a vicious overvaluation of, or excessive desire for, these goods, or (2b) a vicious lack of concern with, or undervaluation of other goods or harms, and (3) to the degree that this pursuit is aimed at securing these goods for oneself; one is also greedy to the extent that one (4) possesses inappropriate attitudes that manifest a vicious overvaluation of (or excessive desire for) these goods, and (5) acts or would act in relevant conditions on the attitudes in (4).

Take clauses (1) and (2). With respect to clause (1), pursuit of a good can involve devoting time, money, and other resources toward acquisition of a good or set of goods, and would include successful acquisition itself.[3] Clauses (2a) and (2b) focus on the factors that lead to the excessive pursuit described in (1). Intuitively, greed involves irresponsibly overestimating the value of the good pursued, or downplaying/ignoring alternative goods, or the harms associated with pursuing the good.

For example, an agent may tend in general to enjoy material goods of some particular kind, and thus tend to form beliefs that—conveniently enough—suggest that such goods are of greater value than they actually are. Other cases may involve envy, as when an agent sees others possessing a good, and comes to value this good excessively precisely because others possess it. In each case, the greedy overvaluation is a result of a culpably irresponsible judgment of a material good's value. The other possibility is that one ignores or undervalues other goods or harms in an epistemically irresponsible fashion. Thus, out of sloth one might fail to investigate the negative impacts of one's consumption on the environment, or only engage in a superficial assessment. Apathy might lead one to fail to evaluate the value of rival goods, or to ignore important harms associated with pursuing a given material good, leaving one with a comparatively excessive valuation of the material good as a default.[4]

Now take clause (3). Greed involves excessively pursuing goods for oneself. Imagine a person who devotes herself to acquiring material goods, but simply in order to give them all away to friends and charities while she lives in a tiny, rundown apartment. Such an agent may excessively pursue material goods, and underestimate the value of her own well-being, but certainly this does not seem to be a matter of greed. Greed requires self-centeredness.

With respect to the second half of the definition, in clause (4) we are concerned with an agent's tendencies toward inappropriate attitudes that manifest greed, where an attitude toward an object manifests greed to the extent that it is a result of vicious overvaluations of or excessive desires for given goods or sets of goods.[5] Thus a thief in prison might exhibit greed in constantly thinking about various goods he wants, even if he cannot take effective steps to acquire them. One's excessively strong desires for a good might lead to a slight—but still inappropriate—downplaying of the importance of one's community, of helping others, and so on. Greed manifests itself in our attitudes toward other objects, not only the immediate objects of our desires.

The final clause, (5), concerns the extent to which an agent acts on the attitudes described in (4), or would so act in relevant circumstances. Thus we might have two greedy agents who have come to resent demands of charitable giving, given their overvaluation of some material good. One agent still gives, even if grudgingly. But the other begins to reduce or even stop giving—and would do so in a wide range of similar circumstances. It seems that the latter agent is greedier than the former. Or similarly, agents could be greedy to the extent that they resent paying taxes (insofar as this detracts from their excessive pursuit of various goods)—they are greedier still when this would lead them to cheat on their taxes.

With this in hand, we can characterize an agent's overall greediness as follows:

The extent to which one is *greedy* (overall) is a function of (1) the number of goods or sets of goods with respect to which one is greedy, (2) the depth or extent of one's greed with respect to these goods, and also (3) the extent to which one possesses inappropriate attitudinal dispositions that manifest greed, and (4) the extent to which one acts upon (or would act upon in relevant circumstances) the attitudes in (3).

The first two clauses are straightforward, and capture the idea that one's overall greed is largely a function of one's greed toward various goods. The more goods with respect to which you are greedy, and the greater the depth of your greed for these goods, the greedier you are, overall. The third and fourth clauses reflect the fact that an agent's greed with respect to various goods can manifest itself in attitudes and actions toward rival goods. Again, for example, a person might begin to resent charitable giving because it takes away money that she wishes to use for her excessive pursuits. Finally, being greedy is a matter of degree, and in its weakest forms it is best understood as simply a limitation to one's temperance or overall virtue. In its more severe forms it becomes a full-fledged vice.[6]

To summarize this discussion, we can consider in table 11.1 an extended example of a man who is greedy for an SUV, indicating how each of the criteria (1) through (5) for greed might be satisfied.

Note that all of the preceding factors are matters of degree (and so the man's level of greediness with respect to the SUV would also vary), and that the precise sources and manifestations of his greed could vary; this is simply one example.[7]

Modest Greed

Where does the definition lead us? I hope to a subtler picture of greed, one that illuminates both why the globally wealthy may in fact be greedy even when not obsessive about wealth or status, and why greed will be an especially problematic vice for future generations.

A common assumption about greed is that greedy individuals experience desires that are excessive with respect to their felt intensity. Thus in

Table 11.1

Clause	Example (of a man who is greedy with respect to an SUV)
(1)	He begins working overtime, cutting back on family activities and spending, and so on, in order to pursue the purchase of an SUV.
(2a)	He excessively desires this SUV because he lets himself be swept away by advertisements, and is jealous of his neighbor who recently bought a similar new vehicle.
(2b)	He never really questions whether his family actually needs this new SUV, or whether a smaller, more efficient vehicle would be viable.
(3)	While he sometimes tries to rationalize his desire for the SUV in terms of looking after his family, what ultimately drives him is his desire to be seen in a top-of-the line new vehicle—it's for him.
(4)	He comes to resent spending time with his child when he is exhausted after his long workdays, and questions whether he really needs to contribute "so much" to charities—money that could instead go toward the SUV purchase.
(5)	Finally, he might actually start giving less to charities because he now sees them simply as wasting money, and also spend much less time with his child due to his changed attitudes in (4).

dictionary definitions we find such terms as "inordinate or insatiate," "intense," and "rapacious."[8] And for many obvious instances of greed, this may well be true. Even so, with this assumption it becomes far too easy for most of us to brush worries aside. "*I* care far more about my family than I do about material things, fame, or wealth. I mean sure, they're nice, but I have things in proper perspective. I'm glad I'm not one of these materialistic, greedy people." But, as our definition makes clear, we can desire and strive for what is excessive for us, even if the felt intensity is quite *modest*. We might set ourselves a goal of buying a large house in the suburbs without our desire being aptly described as "intense" or "rapacious." Yet this could still be greedy behavior on our part.

A second common assumption about greed is that it involves desires for, and a pursuit of, objects that are themselves inherently excessively luxurious or expensive. Thus when some think of greedy individuals, they think of people set on SUVs, mansions, jewelry, money, and so on. While the first common assumption posits excessively intense desires, the second assumption posits excessive *objects* of desire. And again, with closer examination we find that this needs to be reconsidered. Peter Wenz writes: "Consumer society cultivates *greed*, the unlimited desire for more. Without greed consumer demand would flag, the economy would slump, and people would lose their jobs. *Avarice*, an inordinate desire for wealth, is implied by greed. People who want more and more of what money can buy desire unlimited amounts of money" (2005, 206).

Wenz presumably does not intend this statement as a strict definition, but still it reflects and reinforces the image of greed as involving unlimited desires for *excessive items*;[9] similarly with his characterization of avaricious people desiring unlimited amounts of money. But greed is typically far more modest than having unlimited wants (in any strong sense) for material goods and wealth; to think in these terms might lead us to overlook most instances of greed. One could be greedy for a yacht and endlessly more, but one could equally be greedy with respect to such things as clothes, shoes, or the latest electronics (and even when the relevant desires are limited or finite).

A person's life can be shaped by excessively pursuing goods that seem quite modest in themselves. Such a pursuit can result in lost opportunities—to instead give the money involved to environmental groups or other organizations, to spend time engaged in other rewarding activities, and so on. That is, the pursuit might not cause harm in itself, but can still be problematic by leading an agent to miss out on much better uses of the same resources. Beyond this, such excessive pursuit often causes harms—the production and disposal of most electronics involve significant amounts of toxic chemicals, there are environmental costs to shipping goods around the world, and so forth.

A person's overall pursuit or desire for material goods can be relevantly excessive even if each individual desire or pursuit of a good (and each good), taken in isolation, seems modest. Consider a middle-class American who regularly drives to the mall, and buys a shirt here, a DVD there, foods imported from around the world, and so on. Each individual desire and pursuit seems mild enough, but when we look at the *cumulative* effect, it is what we would expect of a greedy person. This agent is still having comparatively large impacts on the environment. She is still encouraging ongoing shipping of goods across the world (with the associated carbon emissions and other impacts), devoting more land to growing cash crops (with the associated loss of habitat or local subsistence crops), and so on. Indeed, her environmental impacts could well be worse than those of the person who lives an otherwise ascetic life while saving for a yacht. Much of the consumption of the globally wealthy seems to arise out of such apparently mild desires, often as a mere habit, a diversion.

Notice that people who attempt to be frugal (voluntarily) can often be greedy in this seemingly modest way. They may buy the cheapest goods they can find, even if they do not buy much. While they might not be spending much money and are not concerned with luxuries, they still cannot be bothered with how various products came to be so cheap. Some of them may suspect that workers have been exploited in some developing nation, or that poor environmental practices were involved, but this does not stop them; they may never investigate fully, or instead conveniently downplay their impacts. They are more concerned with acquiring goods cheaply than with moral or other concerns. *This is a form of greed.* Of course this is not to hold that those who cannot afford better alternatives are necessarily greedy (i.e., there is a difference between a voluntary cheapness and an imposed poverty).[10] Rather, and particularly for those who are wealthy, one must take responsibility for one's purchases and investments, and the place that they hold in one's life. If we focus only on instances of greed involving intensely felt desires for extreme wealth or obviously luxurious material goods we will overlook these far more common—and to that extent, far more troubling—forms of greed.

Must the alternative to such behaviors (and way of life, more broadly) be a life of comparative squalor and misery? We cannot enter into this issue in detail here, but note two important points. First, the citizens of many other wealthy nations have much smaller environmental impacts, and are hardly models of poverty. For example, as of 2009, the total ecological footprint per capita (in global hectares) for the United States has been estimated to be 9.02gha, while for Norway it is 4.20gha, for Germany it is 4.03gha, and for Sweden it is just 2.84gha (Ewing et al. 2009, 57, 73). A significant reduction in environmental impacts for Americans thus seems viable, without a slide into poverty. Second, there is a growing literature in psychology (and related disciplines) suggesting that we tend to significantly overestimate how happy we will be made by acquiring material goods for ourselves, compared to pursuing other goods (see, e.g., Kasser 2002; Cafaro 2005; and Dunn, Aknin, and Norton 2008). To the extent that this literature is correct, a plausible case can be made that those excessively pursuing material goods are less happy than those pursuing other ends, and that a reduced pursuit of material goods need not adversely impact the happiness of the globally wealthy.[11]

Modest Greed in a World of Billions

We have seen that an agent's pursuit of satisfying a set of desires can be greedy, even if the individual desires and pursuits seem quite modest.

These problems are compounded when multiple agents are acting, particularly with respect to justice. To mildly desire a second doughnut at a meeting and to take one may not seem excessive, but the situation changes if there are only a few doughnuts left and many colleagues who have not yet had anything to eat. A person who insists on taking a second doughnut in such circumstances is (barring some unusual justification) being greedy.

Now, note: in terms of felt *intensity*, the desire for a doughnut is likely to be quite mild; and a doughnut seems a modest *object*. But still, for example, if an agent culpably downplays his impacts ("They probably don't want anything; and anyway, other people were taking doughnuts, too"), selfishly inflates his own claim to the good relative to others ("I'm hungry, and it's not my fault they were too slow"), or simply never even so much as thinks of the others (a limiting case of irresponsibly ignoring harms and rival goods while focusing on satisfying our desire for a good), we have an instance of a greedy action.

An obvious analogy emerges here with our actions and our collective impact upon future generations in terms of species loss, resource depletion, using up sinks for carbon emissions, and so on. The cumulative impact of large numbers of people—billions of us—satisfying apparently mild desires for apparently modest goods can be devastating.[12] What may not be greedy in other circumstances (of low overall consumption, low population, and enlightened technology) is now greedy, given current global conditions of growing populations and consumption. Sometimes we culpably downplay our harmful impacts ("I'm sure there are lots of people who are environmentally far worse than I am—they're the real problem"), we selfishly overemphasize our claim to various goods while ignoring the well-being of others ("I've worked for what I've got, and no one has the right to tell me how to live"), and so on; much more common still is simply for no thoughts of impacts on future generations to even cross our minds (see Bendik-Keymer, this volume, on the vice of wantonness). This again qualifies as greedy behavior.

To properly understand and address greed, we need to recognize that we are on a world with billions of other people, and that there will be billions more after us. There are also billions of fellow mammals, birds, and other living things that share this planet with us, and those that will share the planet in the future (See Schlosberg, this volume, for an effort to extend a theory of climate justice to nonhuman entities). This is true for current generations, but will take on still greater importance over decades to come as the human population reaches (and will likely remain close to) its peak on this planet. If each human thinks only in terms of her individual desires and pursuits of goods, each in isolation, she falls into a trap of self-centeredness, one that sustains greed. Instead we must look at our overall patterns of consumption as individuals and as communities within the broader context of a massively consuming human population.

Derek Parfit writes:

> Until this century, most of mankind lived in small communities. What each did could affect only a few others. But conditions have now changed. Each of us can now, in countless ways, affect countless other people. We can have real though small effects on thousands or millions of people. When these effects are widely dispersed, they may be either trivial, or imperceptible. It now makes a great difference whether we continue to believe that we cannot have greatly harmed or benefited others unless there are people with grounds for a serious complaint, or for gratitude. . . . For the sake of small benefits to ourselves, or our families, we may deny others much greater total benefits, or impose on others much greater total harms. We may think this permissible because the effects on each of the others will be either trivial or imperceptible. If this is what we think, what we do will often be much worse for all of us. (1984, 86; cited in Attfield 2009, 229)

Parfit's claims here are of a piece with the current proposal. But notice more explicitly the inverse of what Parfit stresses: when we act, we need to bear in mind that billions of other people are also acting, making trade-offs, and so forth. To act virtuously in a changed and changing world requires awareness of such broader contexts. (Compare Hirsch and Norton, chapter 16, this volume.)

Greed and Collective Action in a Changing World

For all this, we might wonder whether vices such as greed are truly at stake when we examine cases focusing on the presence and behaviors of others. Dale Jamieson raises the issue:

> We should remind ourselves that while a great deal of environmentally destructive human behaviour can rightly be denounced as greedy or vicious, much is humdrum and ordinary. [. . .] Many of our environmental problems have the structure of collective action problems. These involve many people making small contributions to very large problems. They do not intend to cause these problems, and in many cases feel quite powerless to prevent them. The "soccer mom" driving her kids to school, sporting events, and music lessons does not intend to change the climate. Yet, in a small way, that is exactly what she is doing. (2008, 91–92)

There is a worry, then, that many cases that we have treated as involving modest greed might instead be better understood as collective action problems.

The obvious immediate response here is that modest vices and virtues can be relevant to collective action problems. The more people pay attention to such problems, acknowledge their roles in contributing to them, are willing to cooperate, to act as needed, and so on, the better off we will be in trying to address them (compare Thompson, this volume, on a virtue of environmental responsibility). And of course to the extent that our values are skewed, that we downplay the importance of others, that we overvalue material goods, are reluctant to take action (or lack the willpower to do so), and so on, it will be more difficult to address collective action problems.

Still, the worry persists—might our individual actions ultimately be ineffectual, and would this give us reason not to worry about modest greed? For example, should Claire boycott cheap products produced in environmentally unsound fashions, given that her actions alone are hardly likely to cause the manufacturers to cease production of such goods, or even to modify their procedures? Perhaps Claire would simply be making her own life more difficult while contributing almost nothing to solving the problems she is hoping to address; perhaps acting in a modestly greedy fashion would not be so problematic, after all.

There have been several compelling recent discussions of such questions concerning the effectiveness of individual actions, and whether we are obligated to engage in them.[13] Broad themes that emerge from these discussions are that (1) individuals need to engage in social and political activity directed toward solving the problems they are seeking to address—purely private actions will typically have only smaller impacts (e.g., influencing friends and neighbors, perhaps helping to avoid charges of hypocrisy in calling for broader political changes), and (2) it is best for people to have characters that lead to appropriate actions (e.g., reducing misguided consumption, etc.) even if in isolation the individual actions do not seem to have significant impacts; an emphasis on the virtues is warranted.[14]

We can add several important considerations here. First, and perhaps most crucially, note that in morally assessing a person's character and behaviors, far more is relevant than simply the effects of her actions. For example, we can properly negatively assess the character of a racist, even if his attitudes never manifest themselves in overt actions. Or suppose that the racist attempts to promote his views by regularly handing out crude, simple-minded pamphlets that influence almost no one. We do not treat him or his actions as morally nonproblematic even if he has almost no impact in his efforts to promote hatred. Similarly, the character and actions of the mildly greedy can be seen as morally problematic, even if their actions have little overall impact.

Yet we can positively appraise an agent's character and actions, even if the impacts of her actions seem quite modest. We can admire the generosity of a person living in poverty, even if she can only make very small contributions, given her circumstances. We can positively assess those who embrace and exhibit temperance and simplicity, even while their actions as individuals may have little impact on global problems.

To be virtuous requires a mindfulness of the larger context in which we live and act.[15] That certain actions seem common and accepted in our society does not yet show them to be justified. When we reflect upon the demands of justice and benevolence, we see that, for many of us, our pursuit of many goods is excessive and greedy. Virtue requires recognizing changing circumstances; we are now in a world with a rapidly peaking population and growing consumption and waste. We cannot justify our actions by simply declaring that they are widely embraced, and individually seem to have little impact, any more than we can justify the actions and attitudes of a racist, even if his actions in isolation seem to have little adverse impact.

And second: when assessing our possible actions and their impacts, we should focus on those alternatives realistically available to us. For most of our actions there will be only small impacts, but within this range we can make choices. If we consume appropriately, locally grown food, for example, we encourage local producers and cut back on the distance at least some of our food must be transported. One way or another we must eat. But we often exhibit modest greed in buying cheaper products (with dubious origins) when we could viably make wiser, less harmful choices.

Modesty is required in assessing our actions. If we concern ourselves only with actions that are likely to have a direct, significant impact in solving large-scale collective action problems, it might seem that nothing is viable for us as ordinary people. Suppose you were, somehow, to obtain $10 million—an apparently large and significant sum. Even if you were to apply all of this money, it likely would have little direct impact on solving a global problem. Perhaps you could start influencing a few lawmakers in a few countries—but even here, this only goes so far, and it would depend on what other people were doing with other lawmakers. Even the leaders of powerful nations will have only limited influence over other countries and international agreements. With such an approach, only a few people with extraordinary political, economic, or military power would have any reason to concern themselves with their impacts, as they would be the only ones with the direct power to effectively hope to address global collective action problems as individuals. But surely something has gone wrong if our understanding of ethics tells us that only a powerful few bear any responsibility for action in the face of large-scale problems in which we are all implicated, and that the rest of us can revel in greed, ignorance, and complacency as we see fit.

I think these points help to address a worry raised by Jeremy Bendik-Keymer (chapter 13, this volume): that most globally wealthy people seem to be reasonably decent people, yet they are not terribly concerned with the current mass extinction event, or with climate change. Is it too demanding to ask decent people who are already focused on other good projects (and with limited time and money) to address climate change also? This is a difficult question, but the following seems fair. Even given current institutional structures and human limitations (of time, ability, and so on), most globally wealthy people should be more concerned with climate change than they are. While there are limits to what they can do, they can take much better advantage of the opportunities realistically available to them. This might require shifting at least some time and effort away from other projects, but given the potential scale of the harms at stake, a realignment of projects is surely in order. There are many exemplary individuals who are able to pursue a range of worthwhile projects while doing much more than others to attempt to address significant environmental issues. While we cannot know everything and cannot act on behalf of every good cause, there is significant scope for most of us to make better decisions.[16]

More broadly, if we suspect that policy change at national and international levels will take significant time to achieve, we can at least take the local steps available to us. If we wait for change to come from international agreements, we (as individuals) will simply continue to "help" worsen the problems we face for years, in those small ways that are viable to ordinary people as we continue with our modest greed, waiting for change to come. If we worry about crossing various climate thresholds in the decades ahead, it seems foolish to continue to act as we do while waiting for agreements to be reached. Again: we can either try to do our best, or we can rest in modest greed.

Some thinkers will return here to the importance of institutional and political factors while downplaying individual behaviors and character. Both Bendik-Keymer (chapter 13, this volume) and Vogel (chapter 15, this volume) seem tempted to do so. Vogel stresses that, with respect to many environmental issues, individuals are trapped in a "tragedy of the commons" scenario (see Hardin 1968) where any actions they take in isolation will do nothing to solve overall problems; if they consume less, others will consume more, and so on. To be sure, on the one hand, simply having individuals addressing their own modest greed will not be sufficient for solving these problems. But on the other hand, if we do not address the character and actions of individual persons, from where do we expect political or institutional change to come? If there are no exemplars, will others believe that reduced consumption and other green changes are desirable or even viable? There seems to be an assumption that most of us tend to have good values, but cannot act on them viably, given current circumstances. But then why do people vote for politicians who deny the existence of climate change? Why do we fail so often to engage even in easily accessible green behaviors? It is not so obvious that we are simply good people trapped by bad institutions.

The globally wealthy (and others) may mean well, and be decent in some weak sense, but if we are still modestly greedy, apathetic, uninformed, and so on, there is little hope of political change. Even if a group of environmental lobbyists manages to have green legislation passed, if the public does not embrace these values or understand the importance of the issues, they can more than happily elect politicians who will weaken or repeal it. Certainly, institutional change is ultimately needed, but without informed, concerned, and virtuous individuals, such reform is unlikely to occur or to last. Here there can be a spiraling effect—if a large enough group manages to create institutional change, the attitudes of other individuals may change in response; with the new, broader support and change in attitudes, further institutional and political change may become viable, and so on. But individual character and action cannot be dismissed as irrelevant to such change.

A third consideration is relevant: we cannot be perfect, and our circumstances will shape the options open to us; as noted already, trade-offs will be necessary. Suppose a person needs to drive to work, given her particular circumstances. If so, she may need to think about compensating for this—to do so would reflect a recognition of the full costs of her actions (while a modestly greedy person would continue to enjoy the benefits of cheap gasoline while culpably ignoring or downplaying the harms at stake). So she might buy local produce to cut back on other transportation required to sustain her existence, or she might give more time and money to environmental groups. We can allow ourselves some trade-offs in our actions, precisely because the individual impacts are typically so small (though we must keep in mind that billions of others are also acting and making trade-offs). But to be truly decent people, we need to reevaluate our projects and activities from time to time, to ensure that we are helping to address important issues.

And a fourth consideration: how then do we balance all of this—how do we choose acceptable trade-offs without allowing ourselves to slide into modest greed and other vices? Here no simple rules can be given— but one plausible move involves an appeal to our integrity and ideals. Most of us strive to be certain kinds of people; we typically have ideals of compassion, justice, and so on that we can successfully express through our actions. So, for example, while it may be inconvenient to buy environmentally sound goods, there are benefits to such actions—we will have at least some impact, even if minimal, and such actions would be in accordance with our ideals; we could also question whether we even need the goods at all. To ignore our ideals and values as we find them to be inconvenient will both lead to and reflect a lack of integrity and a drift into mild vice; these ideals are typically far more important than many of the other things with which we allow ourselves to become preoccupied in day-to-day life.

Conclusion

Given foreseeable global conditions, we need to shift from thinking of greed in terms of the objects pursued, or even the intensity of the desires for them, to instead looking at the costs of the pursuit in terms of justice and other goods—this is where the excessiveness lies. We each must ask: what are the impacts of my pursuit of various goods in current global conditions—what harms and benefits result, and what could I do otherwise, what alternatives are there? What kind of person and character do my actions reflect? Recognition that now, and for generations to come, each of us will be one among several billion or more humans each shaping the world in small ways, will be crucial to our avoiding greed and other vices. We will need to think in terms of justice, both within and across generations, moving away from a narrow concern with our immediate neighbors and ourselves. It is our thoughtless consumption, our ignoring of alternatives, and our willingness to pass off costs that are now the potentially most worrying drivers of greed.

Acknowledgments

I would like to thank Allen Thompson and Jeremy Bendik-Keymer for their helpful advice throughout this project, and the participants in the Human Flourishing and Restoration in the Age of Global Warming conference held at Clemson University, September 2008, for their feedback; in particular, I am grateful to Phil Cafaro, Ned Hettinger, Martha Nussbaum, Matthew Pianalto, Ron Sandler, Bill Throop, Steven Vogel, Justin Weinberg, and Sarah Wright. Thanks also to Monica and Desmond Kawall, Nixie Knox, and Claire Sigsworth for helpful conversations.

Notes

1. Recent popular books pointing to greed as the source of significant harms include Hamilton and Micklethwait 2006 and Huffington 2004, among many others.

2. I draw this classification from Hurka 2001, chaps. 3 and 4.

3. Note that one could be greedy with respect to a single good (perhaps a mansion), or with respect to a set of goods (perhaps more and more shoes, or home electronics).

4. How would Kawall understand what Vogel (chapter 15, this volume) calls "alienation"? Is it the result of an "epistemic vice"?—eds.

5. What if one were to feel ashamed of one's excessive desires for some material good—would this qualify as an attitude that manifests greed? We can distinguish between attitudes that manifest greed and those that are reactions to it; one's shame would be a reaction to greed, rather than a manifestation.

6. We need not provide a firm dividing line for current purposes. A further question that cannot be addressed fully here concerns how the value of various goods for an agent is determined. But notice that on any plausible theory of value there will be a potential gap between how an agent currently values and pursues an object, and the value that she ought to attribute to it. So long as there is such a possibility of error on the part of an agent, there is a possibility of greed on the current account.

7. Do these nuanced and expanded criteria of greed explain the structural wantonness or inaction that Bendik-Keymer (chapter 13, this volume) or Vogel (chapter 15, this volume) designate?—eds.

8. Take, for example, the OED's entry on "greedy": "1. Having an intense desire or inordinate appetite for food or drink; ravenous, voracious, gluttonous. 2. Eager for gain, wealth, and the like; avaricious, covetous, rapacious" (*The Oxford English Dictionary*, 2d ed., s.v. "greedy"). Or on "greed": "Inordinate or insatiate longing, esp. for wealth; avaricious or covetous desire" (ibid., s.v. "greed").

9. In judging whether an object is excessive, considerations of justice often play a key role. For example, a king's building of a palace is excessive when his subjects are starving, and the resources could instead have been used to improve their lives. Compare this comment with the implications of internalizing ecological integrity into justice as Schlosberg (chapter 8, this volume) proposes.

10. On the other hand, poverty does not preclude the possibility of greed—an agent living in poverty could still exploit others, or overly value and pursue goods that are best seen as luxuries in her circumstances.

11. Note that Kasser (2002) includes the pursuit of fame and of a good image in his overall account of material goods. On the other hand, many of the studies that he discusses focus on material goods or wealth, more narrowly construed (as is our focus here), and provide a plausible basis for the claims just presented.

12. Of course in the original doughnut case the results are far less devastating, but we still arrive at an unjust distribution of goods, and an unjustified thwarting of desires (i.e., the colleagues' desires for something to eat).

13. For example, see Iris Marion Young's work (2004) on a conception of political responsibility and the discussion of her view in Thompson, chapter 10, this volume.—eds.

14. See Jamieson, chapter 9, this volume; Johnson 2003; and Sandler 2010. For discussions focusing on vegetarianism, see, for example, Nobis 2002 and Hudson 1993.

15. Consider the issues of virtue in context discussed in the introduction and especially in part IV, both of this volume.—eds.

16. I attempt to consider some of these issues—of how to divide our attention and information gathering across morally important issues—in Kawall 2010.

References

Attfield, R. 2009. Mediated Responsibilities, Global Warming, and the Scope of Ethics. *Journal of Social Philosophy* 40 (2): 225–236.

Cafaro, P. 2005. Gluttony, Arrogance, Greed, and Apathy: An Exploration of Environmental Vice. In *Environmental Virtue Ethics*, ed. R. Sandler and P. Cafaro, 135–158. Lanham, MD: Rowman & Littlefield.

Dunn, E., L. Aknin, and M. Norton. 2008. Spending Money on Others Promotes Happiness. *Science* 319 (5870): 1687–1688.

Ewing, B., S. Goldfinger, A. Oursler, A. Reed, D. Moore, and M. Wackernagel. 2009. *The Ecological Footprint Atlas 2009*. Oakland, CA: Global Footprint Network.

Hamilton, S., and A. Micklethwait. 2006. *Greed and Corporate Failure: The Lessons from Recent Disasters*. New York: Palgrave MacMillan.

Hardin, G. 1968. The Tragedy of the Commons. *Science* 162 (3859): 1243–1248.

Hudson, H. 1993. Collective Responsibility and Moral Vegetarianism. *Journal of Social Philosophy* 24: 89–104.

Huffington, A. 2004. *Pigs at the Trough: How Corporate Greed and Political Corruption Are Undermining America.* New York: Three Rivers Press.

Hurka, T. 2001. *Virtue, Vice, and Value.* New York: Oxford University Press.

Jamieson, D. 2008. *Ethics and the Environment: An Introduction.* New York: Cambridge University Press.

Johnson, B. 2003. Ethical Obligations in a Tragedy of the Commons. *Environmental Values* 12: 271–287.

Kasser, T. 2002. *The High Price of Materialism.* Cambridge, MA: MIT Press.

Kawall, J. 2010. The Epistemic Demands of Environmental Virtue. *Journal of Agricultural & Environmental Ethics* 23: 109–128.

Nobis, N. 2002. Vegetarianism and Virtue: Does Consequentialism Demand Too Little? *Social Theory and Practice* 28: 135–156.

Parfit, D. 1984. *Reasons and Persons.* Oxford, UK: Clarendon Press.

Sandler, R. 2010. Ethical Theory and the Problem of Inconsequentialism: Why Environmental Ethicists Should Be Virtue-Oriented Ethicists. *Journal of Agricultural & Environmental Ethics* 23: 167–183.

Wenz, P. 2005. Synergistic Environmental Virtues: Consumerism and Human Flourishing. In *Environmental Virtue Ethics*, ed. R. Sandler and P. Cafaro, 197–213. Lanham, MD: Rowman & Littlefield.

Young, I. M. 2004. Responsibility and Global Labor Justice. *Journal of Political Philosophy* 12 (4): 365–388.

12

Are We the Scum of the Earth? Climate Change, Geoengineering, and Humanity's Challenge

Stephen M. Gardiner

So is the answer to a known and increasing human influence on climate an ever more elaborate system to control the climate? Or should the person rocking the boat *just sit down?*[1]
—Gavin Schmidt, "Geoengineering in Vogue"

According to many august scientific reports and bodies, humanity is currently causing environmental change at an unprecedented rate and on a global scale. Moreover, the magnitude of this change is profound. Human activities, we are told, are already ecologically unsustainable; but if current trends continue they will fast become dramatically so. Massive devastation will almost certainly be inflicted on nonhuman life; and there are strong reasons to believe that future humans and the current poor will also suffer severely.

Despite the evidence pouring forth from the sciences, so far not much has been done to address the threat. Instead, we seem in the grip of a profound political inertia (Gardiner 2006; Shepherd et al. 2009). It is difficult to account for this. Nevertheless, its presence seems clearly subject to moral criticism, and in a number of ways (Gardiner et al. 2010).[2] One line of censure focuses on the direct effects of our behavior on other humans and members of other species. For example, it is often pointed out that lack of action imposes harms (or risks of harms) on others, promotes pain and misery, and so on (e.g., Shue 2010). A second avenue of criticism centers on the unfairness or inequities of our behavior. For example, the inertia seems to involve some groups using the power of their privileged sociopolitical and temporal positions to take advantage of those who are physically and temporally distant, and so especially vulnerable (e.g., Gardiner 2011a; Nolt 2011). Both approaches represent important sources of moral concern that need to be discussed. However, this chapter aims to direct our attention to a further component of mainstream moral thinking that also seems relevant. Sometimes we are concerned not only, or even primarily, with what is done to others, but also with what such behavior shows about us. Ethics is not only about our relationships with other morally important entities, it is also about *who we are.*

Theoretical discussions of the environmental crisis have tended to neglect the 'Who—ethically speaking—are we?' question.³ This is partly because of the relative neglect of ethics more generally. But it may also be because even within moral and political philosophy, there is a long-standing tendency to marginalize concerns of character and virtue. Of course, the last forty years or so have seen a significant backlash against this tendency, and one which has resulted in a renewed status for virtue ethics as one of the "top three" normative approaches of our time (with utilitarianism and Kantian theories).⁴ Still, it remains true that contemporary discussion is dominated by more standard vocabularies. This is especially so both in practical ethics in general, and where the specific problem to be addressed involves large collectives, such as communities, countries, generations, and humanity as such.

This chapter makes an initial case for the relevance of the 'Who are we?' question to the topic of global environmental change. In doing so, it does not seek to eliminate or marginalize other ways of talking about these matters. Instead, the aim is to add to (and perhaps reframe) what has already been said from these perspectives, rather than to replace them. Interestingly, the discussion will also suggest that attention to the 'Who are we?' question may facilitate much-needed progress on action.

The chapter focuses on climate change in general, and the recent resurgence of interest in geoengineering in particular. Its main claims are as follows. First, climate change poses a basic challenge to humanity that it has thus far failed to meet. Second, this failure is subject to negative ethical evaluation: it may morally tarnish the lives of those who are party to it, and perhaps irredeemably so. Third, even a moderately successful push toward geoengineering may not be sufficient to blunt this concern. Even if we make the highly optimistic (some would say 'heroic') assumption that intentional manipulation of the climate system might turn out relatively well (e.g., because it is at least a "lesser evil" than allowing climate change to unfold unchecked), the choice to pursue it may nevertheless tarnish the lives of those who make it, and of those who make the final decision to deploy. Fourth, this concern may have motivational efficacy. Moral tarnishing is something that we usually have good reason to avoid. Hence, drawing attention to its relevance might turn out to be useful in overcoming our current political inertia.

The chapter proceeds as follows. The first section places concern about who we are in the context of more usual ethical complaints about global environmental change. The second introduces one way of looking at our current predicament from which the concern naturally emerges. Humanity, it is said, faces a basic evolutionary challenge. Thus far, it seems to be failing that challenge, and in a way that reflects badly on us. The third section discusses some categories of negative ethical appraisal to which failure makes us susceptible, including those involving tarnishing, marring, and blighting evils. The fourth section clarifies the moral presuppositions of the challenge through responses to some basic objections.

The Usual Story

In the realm of climate policy, there seems little doubt that humanity as a whole is in the grip of profound political inertia. By 1994, almost all of the world's nations—including all of the big polluters—had committed themselves to the United Nations' Framework Convention on Climate Change, and so to the objective of "avoiding dangerous climate change." However, since then substantive progress has been elusive. Though significant promises to act have been made at various times by major governments, almost none have been kept, and the overall situation is little short of dire. Globally, emissions have risen dramatically (by around 30 percent at this point), and prior to the recent financial crisis were increasing very quickly (2–3 percent per year), a rate beyond the highest projections of twenty years ago. Domestically, growth has been occurring at substantial rates in most of the large polluters. For example, China has recently surpassed the United States in absolute annual emissions, and is up by more than 150 percent since 1990. Moreover, the United States itself is up by almost 20 percent over the same period, despite beginning with per capita levels more than five times higher than China's (Boden, Marland, and Andreas 2009).

Political inertia is often said to pose two central ethical problems. The first is that failure to adequately address climate change imposes serious risks of severe (even potentially catastrophic) harms and/or other costs on innocent and vulnerable others. Hence, it appears to violate a particularly weighty kind of moral constraint.

The second is that inertia brings on major injustices. To begin with, there are issues of global justice. Most notably, projections of the likely geographical distribution of impacts in the short- to medium-term strongly suggest that poorer nations, and poor people more generally, will bear the brunt. This is a problem both in itself, and also because the richer nations, and rich people more generally, are responsible for much bigger contributions to the problem, both historically and at current levels of emissions. Hence, the inaction of the rich appears to amount to taking advantage of the poor and more vulnerable, and especially their vastly inferior bargaining position.

Similarly, there are issues of intergenerational justice. Climate change is subject to large time lags, such that it is highly likely that the worst impacts are substantially deferred. This strongly implies that the current generation is taking advantage of its privileged temporal position with respect to its successors in indulging in political inertia.

Finally, there are concerns about ecological injustice. Many of the impacts of climate change are likely to be directly inflicted on nonhuman animals, plants, and ecosystems, or else passed on to them by human adaptation and coping efforts. Hence, there are concerns about humans taking advantage of their ability to pass on the costs of their behavior to other species.[5]

In my view, concerns about harm and injustice are apt. Still, we can enrich our account of the ethical problem posed by inertia if we also call attention to the question of who we are, ethically speaking. As an illustration, consider three charges that could be made against the current generation of the world's affluent, each of which tends to both accommodate and extend the more standard criticisms.[6]

The first charge is one of *recklessness*. Our lack of appropriate action imposes risks on the vulnerable that are not just severe (the usual criticism), but also at least seriously unjustified, and perhaps deeply thoughtless and wanton (see also Bendik-Keymer, chapter 13, this volume). The charge of recklessness thus extends the focus of criticism beyond the risk itself to the issue of who is imposing it, on whom, under what circumstances, and for what purpose. One effect of this is to help address some potential weaknesses of infliction of harm (and cost) arguments. For example, these are sometimes criticized on the grounds that the harm is uncertain, and so may not be realized. This is already not a very strong criticism, since the extent of scientific uncertainty is often exaggerated, since some kinds of uncertainty are central to the practical challenge, and since infliction of severe risk on others is usually a moral issue regardless of whether things ultimately turn out well for other reasons (such as luck). But the charge of recklessness clarifies and reinforces this response. If we ask what kind of person (or community, nation, or generation) would impose risks of severe climate harms on others under our current epistemic circumstances, we cast the issue in a new, and starkly unflattering, light (see, e.g., Gardiner 2011a).

The second charge against us is one of *callousness*. The extent of our recklessness strongly suggests a profound indifference to the concerns of those who must reap the consequences of our behavior. This accusation incorporates, but also enriches, the standard complaint of injustice. For example, that charge is sometimes disputed on the grounds that the correct accounts of global and intergenerational justice are matters of controversy. This objection is already dubious given the strong ethical consensus surrounding the claim that the behavior of the current generation of the affluent is seriously unjust (e.g., Singer 2002; Shue 1999; Gardiner 2011a). However, the charge of callousness puts the issue into a broader perspective. If we are unwilling even to consider our potential victims in our deliberations (or at least take their concerns seriously), then a core feature of the problem is that for us the question of justice is not even really on the table. Such callousness signals a profound resistance to this central ethical value that is even more morally troubling than normal injustice.

The third charge is that we may be rebuked for our apparent *shallowness*. There is some plausibility to the complaint that much of our failure is itself caused by a commitment to lesser values. To begin with, the failure seems to involve a lack of social and political engagement. Politicians, we are told, are more concerned with reelection than with saving the planet, and many citizens are more concerned about the latest celebrity scandal than climate catastrophe. More generally, there is a suspicion that humanity is largely reliant on institutions that are good at promoting short-term and sometimes fairly superficial concerns, but not so good at furthering long-term values with intergenerational and ecological dimensions.

More fundamentally, there are worries about the underlying causes of this malaise. Most obviously, it will be said that the harms inflicted and injustices perpetrated by the present on the future are done out of narrow self-interest without any respect for ethical concerns. This may be damning enough. However, a further concern is in some ways even more disturbing. Perhaps the benefits whose pursuit brings on the threat of environmental catastrophe ultimately are at best relatively unimportant, and at worst close to trivial, *even to us*, and especially in comparison to the damage done. On this view, the ultimate source of the problem is not a deep conflict between self-interest and the environment, and our opposition to "self-sacrifice" understood in any robust sense. Instead, it results from an overly strong commitment to what are at best minor values, and at worst meaningless fluff.[7]

To make this thought vivid, consider complaints like the following: "Is the current generation so committed to 'soft pleasures' (such as large cars that usually carry only one person, or enormous homes maintained at 71°F year round) that it is willing to inflict real catastrophe on the future in order to be sure of keeping them?"; "Is it so deluded about what matters that it will remorselessly pursue a high-consumption lifestyle even when, in its heart of hearts, it does not find any real fulfillment in this pursuit?"; "Is it really willing to inflict famine, disease, and death on others just so its members can continue working too much, eating too much, and being too stressed out?" If such complaints resonate, then our ethical failure does seem to be in large part one of knowing who we are or aspire to be.

The three charges just introduced (recklessness, callousness, shallowness) illustrate how the ethics of who we are might have something to contribute to our understanding of global environmental tragedy. Still, we might wonder whether such thoughts also point toward a more general evaluative framework. It is to this issue that I now turn.

Humanity's Challenge

Extending the standard criticisms of environmental behavior in the direction of the 'Who are we?' question in part involves a shift from a largely "victim-centered" to a more "agent-centered" evaluative focus. Given this, it is natural to ask: how should we understand the perspective of the relevant agents as a matter of ethics? Presumably, there is more than one possible answer to this question, and rival answers must be considered and assessed. In this section, I sketch an initially plausible framing grounded in what I take to be fairly weak and widely shared ethical intuitions about the problem.

Let us begin with two claims. The first concerns our basic situation. It states that humanity is, in geological and evolutionary terms, a recent arrival on the planet, and is currently undergoing an amazingly rapid expansion, in terms of sheer population size, technological capabilities, and environmental impact. The second claim concerns the implications for our agency. It asserts that, as a result of the first claim, a basic question facing us as humans is whether we can meet the challenge of adapting to the planet on which we live. Together, these two claims embody and reflect a specific normative perspective on our predicament. They identify certain features of our complex circumstances as salient, and propose that these features conspire to pose a question to us, considered together as humans, that we can succeed or fail at answering. Let us call this perspective *Humanity's Challenge* (or simply '*the Challenge*').[8]

The science mentioned at the outset of the chapter suggests that humanity is currently failing to meet *the Challenge*. In the case of climate change, the failure seems severe. Indeed, there the problem of political inertia is so pronounced—and the threat of continued failure so palpable—that the claim is now being made that we must prepare to attempt intentional manipulation of the climate system on a planetary scale. Such "geoengineering" is widely recognized to be environmentally and politically risky. Nonetheless, we are told, preparations should begin, because ultimately we may find ourselves in a nightmare scenario where we are forced to choose geoengineering as "the lesser evil" (Crutzen 2006; Victor et al. 2009; cf. Gardiner 2010).

If the move toward geoengineering represents a failure, is it an ethical one? Not all failures of *Humanity's Challenge* need be so. One can imagine cases were a group fails to meet the challenge for reasons of brute bad luck, for example; or because of excusable logistical or technological errors.[9] Nevertheless, in this case, there is good reason to believe that the failure is at least partly ethical, involving both moral and political elements. Let us begin with two motivating thoughts.

First, there is widespread evidence of political failure. Consider, for example, the remarks of Connie Hedegaard (then the Danish Minister for Climate and Energy, and now the European Commissioner for Climate Action) two months before the widely derided Copenhagen meeting in 2009: "If the whole world comes to Copenhagen and leaves without making the needed political agreement [on mitigation and adaptation], then I think it's a failure that is not just about climate. Then it's the whole global democratic system not being able to deliver results in one of the defining challenges of our century. And that . . . should not be a possibility" (Von Bulow 2009). In light of the subsequent failure of Copenhagen, Hedegaard's remarks suggest that existing political institutions are not functioning as they need to in order to solve a fundamental social problem of profound importance, and that this raises worrying questions about their philosophical limitations and justification (see Gardiner 2011a, 2011c).

Second, the slide toward geoengineering also seems to involve a moral failure. This can be made more vivid through the invocation of a heuristic. Call this *Fouling the Nest*:

As a species, we began with a perfectly serviceable planet to live on, but now we have undermined that. We have, in elementary terms, "fouled the nest." We could clean it up—that would be the most direct approach, the one most likely to work—but so intent are we on continuing our messy habits, that we will pursue any means to avoid that, even those that impose huge risks on others and involve further alienation from nature. (Gardiner 2010)

Fouling the Nest is, I take it, a recognizable perspective, resonating with concerns often raised about our current predicament, especially from within the contemporary environmental movement, but also from outside of it. Arguably, in light of that perspective, the decision to pursue geoengineering has a certain salience. It constitutes the crossing of a new threshold on the spectrum of environmental recklessness—attempting intentional manipulation of the global system—and provides fresh and further evidence of just how far we are willing to go. If so, such a decision is hardly neutral. Instead, it seems to embody a recognition of continued and deepening failure. This is registered when some people express horror at the possibility of geoengineering by saying, "Has it really come to this?"

Such thoughts are natural on a perspective that highlights the "Who are we?" question. If pursuing geoengineering constitutes some kind of ethical failure, much might be said about the risks of harm and misery to which it exposes other people and species, and more might be added about the extent to which the imposition of such risk is unfair and unjust to various individuals and populations (human and nonhuman). Still, it is also true that failure would have implications for how we think about those who are complicit in it. Is it not natural to ask: "Who are these people? How could they do such a thing?" Indeed, this may well be our first and overwhelming moral reaction to the *Fouling the Nest* scenario.

These concerns resonate with some of the features of the *Fouling the Nest* scenario. The sense of the scenario is that the failure is *self-inflicted*, that it is in some sense *immature* (even infantile), that it demonstrates *a lack of even minimal proficiency* in a basic task, and that it does so despite the presence of a strong negative feedback mechanism. Moreover, the motivating idea of the scenario is that the failure is a basic and also somewhat pathetic one. It reveals a fundamental incompetency in agency.

Initially, the *Fouling the Nest* scenario may seem more than a little extreme, even as a heuristic. (Surely such talk is the product of environmentalist hysteria!) However, elements of it seem thoroughly mainstream. For example, many internationally renowned climate scientists use similar analogies in describing their own reservations about geoengineering. Stephen Schneider of Stanford, for example, likens the climate change problem to heroin addiction, and compares the decision to pursue geoengineering to choosing "a massive substitution of [planetary] methadone" over "slowly and surely" weaning the addict (Schneider 1996, 299–300). Hence, he treats humanity as suffering from diminished agency, and as choosing unwisely because of it.

Similarly, Gavin Schmidt voices his reservations about geoengineering by suggesting that there is something absurd about its pursuit. Schmidt offers the analogy of a small boat being deliberately and dangerously rocked by one of its passengers. Another traveler offers to use his knowledge of chaotic dynamics to try to counterbalance the first, but admits that he needs huge informational resources to do so, cannot guarantee success, and may make things worse. Schmidt asks: "So is the answer to a known and increasing human influence on climate an ever more elaborate system to control the climate? *Or should the person rocking the boat just sit down?*" (Schmidt 2006; emphasis added). (Call this scenario *Rocking the Boat.*) The rhetorical nature of Schmidt's question presumably underlies the emotional force of the example. What kind of person rocks the boat and then refuses to sit down? Isn't this a reckless, callous, and shallow individual?[10] Isn't this also like fouling the nest?

Schmidt's example provides further evidence for the idea that not all ethical concerns about climate change and geoengineering are straightforwardly reducible to the standard objections of harm and injustice. Notice that there is an important disanalogy between *Fouling the Nest* and *Rocking the Boat*: *Fouling the Nest* need not involve harms or injustice to other beings; but Schmidt's *Rocking the Boat* most naturally does. This is interesting, because *Rocking the Boat* seems to add to the moral force of *Fouling the Nest*, rather than supplying or replacing it. *Fouling the Nest* has serious force of its own.[11]

This point might be pushed further to suggest that the role of the usual concerns about harm and injustice is to help characterize some ways in which failure to meet *Humanity's Challenge* is manifest, rather than capturing the core failure itself. They represent, if you like, an "internal" perspective on failure. But we might also want an "external" perspective, one that describes the character of the problem from the outside, in its most general terms. Plausibly, this rests on an understanding of *Humanity's Challenge* as such, and frames the internal perspectives. After all, it is far from clear that global environmental change that did not cause harm or bring on injustice would, just in virtue of that, be compatible with meeting the challenge. Surely wider, "external" questions could be asked. Suppose, for example, that humanity could render the Earth lifeless, retreating to artificial domes on the surface, without committing other kinds of injustice (e.g., because they take the animals and plants with them). Still, this seems to involve some form of *Fouling the Nest*, and at least a partial failure of *Humanity's Challenge*.

Tarnishing

The next section addresses four initial objections to *Humanity's Challenge*. However before turning to these, let us first consider how the negative evaluation brought on by failure might be understood, and why a desire to avoid it may be motivating.

People's lives are often subject to serious negative evaluation, and in different ways. The most obvious instances of this involve seriously evil actions, such as those of premeditated murder or genocide. Committing such actions prompts a negative ethical evaluation of a person's life that is deep and serious. We might say that it stains or tarnishes them from a moral point of view.

To this we might add that such evils are sometimes irredeemable, in the sense that they cannot be expunged or even perhaps outweighed (e.g., by other good actions). In such cases, we might speak of the evil as "blighting." It leaves a permanent negative mark on the agent's life, perhaps even one that makes it impossible for a positive (or neutral) overall evaluation to be restored.

We might continue by pointing out that some believe, more controversially, that a person's life can be compromised by circumstances beyond their control even if they themselves make no evil choices. Aristotle, for example, claims that even though Priam of Troy was virtuous, his life was not a flourishing one. The tragedy that befell his family and city in the Trojan War was sufficient to undermine that claim.[12]

Finally, and even more controversially, we might acknowledge a more ambiguous form of tarnishing. This occurs when an agent does the best she can in a situation, but what she does remains in some sense evil. For instance, suppose that the agent finds herself in a morally tragic situation where any choice opens her up to serious negative moral assessment, so that even if she (justifiably) chooses the "lesser evil," her life is stained (and perhaps even blighted) by the choice. (A classic example is given in William Styron's novel *Sophie's Choice*.)[13] We can call such special cases one's of "marring evil." They involve a negative moral evaluation of an agent's action (or actions) that is licensed when the agent (justifiably) chooses the lesser evil in a morally tragic situation, and which results in a serious negative moral assessment of that agent's life considered as a whole.[14]

Such classifications are relevant to global environmental tragedy in a wide variety of ways. Most obviously, those who fail as political leaders risk tarnishing, and indeed blighting, their own careers and lives. This seems highly plausible. For example, in the long run, who will dwell on the short-term economic and strategic achievements of any president or prime minister who fails to take serious steps to prevent a highly predictable catastrophe for their own countries and for humanity as such? Such leaders will be derided by history as surely as Neville Chamberlain usually is for appeasement and his "Peace for Our Time" speech.

Less obviously, the slow drift toward geoengineering may bring on instances of marring. In particular, those who deploy geoengineering may end up inflicting grave harms on innocents that may otherwise not have occurred, such that the deployment mars the choosers' lives, even if it is the best ("lesser evil") choice available. Suppose, for example, that geoengineering really does cause less harm overall than climate catastrophe, but the harm that is left accrues to different individuals. For instance, suppose that it stabilizes temperatures on most of the planet, but dramatically slows the Indian monsoon, with impacts that are much worse for those affected than the temperature rise would have been (Robock, Oman, and Stenchikov 2008). In such cases, those who deploy harm innocents through their agency, and this may be a marring evil (even if it is a "lesser" evil overall). One can certainly imagine the infliction of a devastating famine on India being something that people would find, as the expression goes, "hard to live with." Indeed, this is a prominent feature of other marring cases.[15]

Most interestingly of all, it may be that those who bring about such a nightmare choice situation, say by refusing to limit emissions earlier, are guilty of a special kind of unethical behavior. They not only bring on the nightmare itself, but also are responsible for handing a marring choice to others. This is a kind of wrong that may be hard to account for in ethical terms that do not invoke the 'Who are we?' question. It may also amount to an independently marring evil that blights those who bring it about.

Such points suggest that concerns about tarnishing are highly relevant to climate change policy in general and geoengineering more specifically. Moreover, this may make a difference to moral motivation. Not only are tarnishing, blighting, and marring serious matters, but they accrue directly to those who commit wrongs rather than to their victims. Hence, they have the potential to exert a strong pull on agents.

Objections and Replies

Let us now clarify and extend the position just presented through responses to four basic objections.

Objection 1: Too Grand!

The first questions the applicability of the analysis. Surely, it will be said, *Humanity's Challenge* is too grandiose and unworldly a model. For one thing, it presupposes a unity to the human species that does not and may never exist, and endorses an evaluative perspective that many do not accept, and some would actively resist. (Call this 'the unity complaint.') For another, the analysis is insensitive to the fact that most individuals, and indeed most nations on the planet, have little or no ability to influence global environmental policy in any meaningful way. Most obviously, billions live in poverty and in countries with no geopolitical power. However, it is also true that many working- and middle-class citizens of the richer countries exert little or no influence over their nations' policies, especially by comparison to large corporations, the socioeconomic elites, and so on. How then are they implicated in the general failure? Indeed, would it not be deeply unfair and unjust to suppose that they are? (Call this 'the absence of power complaint.')

The grandiosity objection has some truth to it, and each of the complaints raises serious philosophical questions that cannot be confronted here (see, e.g., Young 2010). Still, we should not be too quick to dismiss *Humanity's Challenge* on these grounds. In particular, according to a long tradition in political theory, political institutions and their leaders have legitimacy to the extent that citizens delegate their own responsibilities and powers to them, so that they can act in the name of the citizens in order to solve problems that either cannot be addressed, or else would be poorly handled, at the individual level.[16] From this simple model, several points follow (see also Gardiner 2011b).

To begin with, offhand the model implies that the most direct responsibility for failure falls on recent leaders and current institutions. If authority is delegated to them to deal with serious social problems—including global environmental problems—then they are failing to discharge the relevant responsibilities and are subject to moral criticism for this.

Of course, it is plausible to argue that existing institutions were simply not designed to deal with large global and intergenerational problems, let alone *Humanity's Challenge* more generally; hence, perhaps they also lack the unity and power necessary for the task, and the assignment of responsibility to them is also unfair. There is some truth to this claim, and we will return to it in a moment. Nevertheless, we should not concede the point too quickly. After all, many existing leaders and institutions have not been slow to take up the issues and assume the mantel of responsibility, making many fine speeches, organizing frequent meetings, promising progress, making the topic a campaign issue, and so on. Hence, even if this role was not originally envisioned, many actors have acted as if it does belong to them. Hence, they can be held responsible for that, and for their subsequent failure to deliver. They can hardly claim to be ignorant of, or to have refused, the responsibility.

Nevertheless, the more important issue is the following. Suppose that humanity does lack the appropriate institutions to deal with global environmental change. What follows? If political institutions normally operate under delegated authority from citizens, then the answer seems clear. This is a case where the delegation has either not happened, or else has failed to be successful. Under such circumstances, it is natural to claim that the responsibility falls back on the citizens either to solve the problems themselves, or else, if this is not possible, to create new institutions to do the job. If they fail to do so, then they may be subject to moral criticism, for having failed to discharge their responsibilities.[17]

Now, of course, this response may invite a third version of the grandiosity objection, that it is ridiculously demanding, and indeed utterly "unworldly." ("Get real! How could such widespread political failure by the great and powerful possibly fall back on you and me?") Nevertheless, this complaint ignores the fact that the basic move being made is an utterly standard and traditional one in mainstream political theory, often made (for example) in discussions of rights to civil disobedience and revolution. Given this, *Humanity's Challenge* need not imply an unusual model of political responsibility, or one that faces insurmountable obstacles relative to prevailing views.

In conclusion, though it may be true that the political unity and allocation of powers among human beings that is necessary for a good response to *Humanity's Challenge* does not yet exist, this need not undermine the ethical claim that it should. After all, solving other pressing political problems often seems to require political institutions that when absent are difficult to achieve, and that, other things being equal, some (or most) would rather do without. (Think of examples such as endowing the state with a monopoly over the legitimate use of force, or the setting of rules of property.) Still, if such institutions are necessary to solve fundamental political problems, then other things are not equal.

Objection 2: Too Remote
The second objection to the *Humanity's Challenge* analysis is that concerns about tarnishing, especially at the level of humanity itself, are just too remote to have motivational, and perhaps justificatory, import. Consider the following suggestion (Thompson 2009). In normal cases of tarnishing, like Sophie's choice, there is a clear psychological effect (usually pain and suffering) caused by the evil done and this has motivational implication. However, the central actors in the climate case are collectives (e.g., communities, nation-states, generations, and humanity as such). Since they do not have psychological states, and since the connection between their actions and the mental states of their members is unclear, what, then, is the motivational force of the prospect of their becoming tarnished?

One part of the answer relies on an appeal to experience. In the real world, many people do find their lives ethically compromised by their connections to various groups. In the recent past, some have been ashamed to be white South Africans during the time of apartheid, or Germans during the time of the Nazis. Moreover, for many this shame persisted even when they not only played no direct role in the evil activities of these regimes, but actively resisted them as well. One advantage of the delegated authority model is that it helps to account for this. It suggests that those who resist evil regimes that act in their name do feel morally compromised by their failure to delegate their authority to solve moral problems. The tarnishing can be real even if is also blameless, as in Aristotle's example of Priam.

Objection 3: Overwhelmed by Context

The third objection claims that concerns about tarnishing in the geoengineering case are largely overwhelmed by the wider context. Specifically, it is said that the decision to pursue geoengineering will be made amid such spectacular political inertia that the agents involved (nations, generations, and so on) are already so substantially, and perhaps irredeemably, tarnished by their behavior that any further blighting is likely to have minimal force (Preston 2009).

This objection has some pull. Still, resistance to a single blighting evil seems to me often admirable even in cases where there is no prospect of redeeming a life already severely blighted. Think of Darth Vader's final turn away from the Dark Side in *Return of the Jedi*. It cannot really overcome years of atrocities, but it is something, and this seems to motivate him.

Objection 4: Can't Motivate in Marring Cases

In addition, there is more to be said about the importance of tarnishing than its motivational force. To see this, consider a fourth and final objection. This maintains that invocations of marring at least are ineffective and inappropriate. "Ought implies can," it is claimed; hence, it is both pointless and misguided to evaluate the behavior of those who have no real choice about what they do. In context, if we must pursue geoengineering anyway, as a "lesser evil," then discussion of these matters is useless, and should be avoided.

My response is threefold. First, we should not be too quick to concede that every so-called lesser evil should be chosen, or that our efforts to do better will be in vain. Evils are, after all, usually to be avoided, and it is a favorite strategy of the corrupt to insist that the evils they wish to pursue are overridingly "necessary."

Second, in context we should be aware that even if geoengineering does turn out to be a lesser evil that must (morally) be done, this has wider ethical implications. In particular, it matters how and why it should be done; and those who are tarnished either by doing it, or making it necessary, may have special responsibilities that are framed by this fact. For example, if an actual intervention did undermine the Indian monsoon, then the agents of geoengineering (e.g., those responsible for political inertia, the eventual geoengineers, etc.) might be morally obliged to aid, comfort, and seek forgiveness from those Indians negatively affected.

Third, beyond these matters, even if it is true that talk of tarnishing makes no difference to action or motivation at all, it may still be worthwhile. Sometimes ethical analysis plays roles other than that of guiding action. In this case, the role might be that of *bearing witness* to what has occurred. This might be important not only for the victims of evil, but also for its perpetrators, and perhaps especially when these are blamelessly tarnished or marred. We might note that in the classic case of Sophie's choice, Sophie herself laments: "In some way I know I should feel no badness over something I done like that. I see that it was—oh, you know—beyond my control, but it is still so terrible to wake up these many mornings with a memory of that, having to live with it. When you add it to all the other bad things I done, it makes everything unbearable. Just unbearable" (Styron 1979, 538). If ethical theory can help Sophie to understand her pain, it plays a worthy role even if this makes no further difference in the world.

Conclusion

This chapter illustrates how concerns about who we are ethically-speaking play an important role in understanding the ethics of global environmental problems, one that adds to (and perhaps reframes) more standard concerns about harm and injustice. It does so through a discussion of political inertia in climate policy, and the looming push toward geoengineering. It suggests that we understand this as involving, at least in part, a failure by humanity to respond to a basic evolutionary challenge, where this failure is in part an ethical one. It also suggests that such failures may tarnish or mar the lives of individuals and collectives, perhaps irredeemably. Finally, it invokes a (traditional) model of delegated political responsibility to explain why the failure and the tarnishing may belong to ordinary citizens as well as to political leaders and institutions. This normative perspective indicates that global environmental change threatens the moral evaluation of our own lives, as well as of our generation, our communities, our nations, and humanity itself. This thought should be motivating. After all, who wants to be the "scum of the earth"?

Acknowledgments

This chapter draws on and substantially extends some remarks from Gardiner 2010.

Notes

1. Emphasis added.

2. Some argue that the message from mainstream science is mistaken. (Many of the rest of us simply hope that it is so.) However, it is far from clear that such skepticism (or hope) justifies the current inertia. If the best evidence we have suggests a realistic possibility of impending catastrophe, and if this evidence passes a high threshold of scientific plausibility, then surely we have reason to act, and especially to consider some kinds of precautionary measures. If we do not, there appears to be a strong case that we are failing in our responsibilities toward the future. Moreover, even if it should turn out that the best science of the time is mistaken, we might still be subject to moral censure for ignoring it. Sometimes being lucky is not enough. (I would not deny that issues of luck infect moral evaluation in interesting ways. See Nagel 1979 and Williams 1976.)

3. Notable exceptions include Jamieson 1992 (revised as chapter 9, this volume), 2007, Sandler and Cafaro 2006, Sandler 2007, and McKinnon 2009, although most of these focus solely on the individual level. The broader concern often shows itself in nonphilosophical discussions (e.g., McKibben 1989).

4. It is, of course, true that not all those who are in sympathy with the backlash call themselves virtue ethicists, and that another result of the backlash has been the renewal of sophisticated discussions of character within the other theories.

5. See Schlosberg, chapter 8, this volume.

6. See also Bendik-Keymer, chapter 13, this volume. For a different kind of illustration, see Thompson, chapter 10, this volume.

7. Consider Kawall's sense of greed (chapter 11, this volume).

8. This need not be the only relevant normative perspective.

9. In addition, one could also imagine cases where the failure is moral, but not of a very serious kind.

10. For example, in context I assume Schmidt intends that the rocker is acting simply for his own amusement, or some other comparatively trivial purpose.

11. One can rock one's own boat, and this could be a failure that is akin to *Fouling the Nest*. But it is not the most salient feature of the usual *Rocking the Boat* scenario, and the 'sit down' complaint.

12. Aristotle, *Nicomachean Ethics*, Book I. 9–10.

13. This negative assessment may arise because her life cannot be seen as flourishing; more controversially, it may also occur because the life now can no longer be counted virtuous.

14. Gardiner 2010. This definition presumably requires refinement, but this is not the place for such work.

15. Consider Williams's famous examples of Jim and the Indians, and George and the chemical warfare job (Smart and Williams 1973).

16. For discussion, see Shockley, chapter 14, this volume.

17. Of course, something depends on the cause of the failure to discharge. Some argue that it is at least partly a failure of values (e.g., see Jamieson, chapter 9,

this volume; Shockley, chapter 14, this volume; Thompson, chapter 10, this volume). Nevertheless, this does not imply that the relevant agents are not responsible (see also Gardiner 2011b).

References

Boden, T., G. Marland, and R. J. Andreas. 2009. "Global CO2 Emissions from Fossil-Fuel Burning, Cement Manufacture, and Gas Flaring: 1751–2006." Carbon Dioxide Information Center, Oak Ridge Laboratory, Oak Ridge, TN. http://cdiac.ornl.gov/trends/emis/overview_2007.html (accessed December 10, 2010).

Crutzen, Paul. 2006. Albedo Enhancement by Stratospheric Sulphur Injections: A Contribution to Resolve a Policy Dilemma? *Climatic Change* 77: 211–219.

Gardiner, Stephen M. 2006. A Perfect Moral Storm: Climate Change, Intergenerational Ethics and the Problem of Moral Corruption. *Environmental Values* 15: 397–413.

Gardiner, Stephen M. 2010. Is "Arming the Future" with Geoengineering Really the Lesser Evil? Some Doubts about the Ethics of Intentionally Manipulating the Climate System. In *Climate Ethics: Essential Readings*, ed. Stephen M. Gardiner, Simon Caney, Dale Jamieson, and Henry Shue, 284–318. Oxford, UK: Oxford University Press.

Gardiner, Stephen M. 2011a. *A Perfect Moral Storm: The Ethical Tragedy of Climate Change*. Oxford, UK: Oxford University Press.

Gardiner, Stephen M. 2011b. Is No One Responsible for Global Environmental Tragedy? Climate Change as a Challenge to Our Ethical Concepts. In *Ethics and Climate Change*, ed. Denis Arnold, 38–59. Cambridge, UK: Cambridge University Press.

Gardiner, Stephen M. 2011c. Rawls and Climate Change: Does Rawlsian Political Philosophy Pass the Global Test? *Critical Review of International Social and Political Philosophy* 14 (2): 125–151. Special Issue on Climate Change and Liberal Priorities, ed. Catriona McKinnon and Gideon Calder.

Gardiner, Stephen M., Simon Caney, Dale Jamieson, and Henry Shue, eds. 2010. *Climate Ethics: Essential Readings*. Oxford, UK: Oxford University Press.

Jamieson, Dale. 1992. Ethics, Public Policy, and Global Warming. *Science, Technology and Human Values* 17 (2): 139–153.

Jamieson, Dale. 2007. When Utilitarians Should Be Virtue Theorists. *Utilitas* 19 (2): 160–183.

McKibben, Bill. 1989. *The End of Nature*. New York: Anchor.

McKinnon, Catriona. 2009. Runaway Climate Change: A Justice-Based Case for Precautions. *Journal of Social Philosophy* 40 (2): 187–203.

Nagel, Thomas. 1979. Moral Luck. In *Mortal Questions*, ed. Thomas Nagel, 24–38. Cambridge, UK: Cambridge University Press.

Nolt, John. 2011. Greenhouse Gas Emission and the Domination of Posterity. In *The Ethics of Climate Change*, ed. Denis Arnold, 60–76. Cambridge, UK: Cambridge University Press.

Preston, Christopher. 2009. Comments on Gardiner's "Should We Arm the Future with Geoengineering?" American Philosophical Association, Pacific Division, Vancouver, BC, April.

Robock, A., L. Oman, and G. L. Stenchikov. 2008. Regional Climate Responses to Geoengineering with Tropical and Arctic SO2 Injections. *Journal of Geophysical Research* 113: D16101.

Sandler, Ronald. 2007. *Character and Environment.* New York: Columbia University Press.

Sandler, Ronald, and Philip Cafaro, eds. 2006. *Environmental Virtue Ethics.* New York: Rowman and Littlefield.

Schmidt, Gavin. 2006. "Geoengineering in Vogue." *Real Climate,* June 28. http://www.realclimate.org/index.php/archives/2006/06/geo-engineering-in-vogue/ (accessed December 10, 2010).

Schneider, Stephen. 1996. Geoengineering: Could—or Should—We Do It? *Climatic Change* 31: 291–302.

Shepherd, John, Ken Caldeira, Peter Cox, Joanna Haigh, David Keith, Brian Launder, Georgina Mace, Gordon MacKerron, John Pyle, Steve Rayner, Catherine Redgwell, and Andrew Watson. 2009. *Geoengineering the Climate: Science, Governance and Uncertainty.* London: Royal Society.

Shue, Henry. 1999. Global Environment and International Inequality. *International Affairs* 75: 531–545.

Shue, Henry. 2010. Deadly Delays, Saving Opportunities: Creating a More Dangerous World? In *Climate Ethics: Essential Readings,* ed. Stephen M. Gardiner, Simon Caney, Dale Jamieson, and Henry Shue, 146–162. Oxford, UK: Oxford University Press.

Singer, Peter. 2002. *One World: The Ethics of Globalization.* New Haven, CT: Yale University Press.

Smart, J. J. C., and Bernard Williams. 1973. *Utilitarianism: For and Against.* Cambridge, UK: Cambridge University Press.

Styron, William. 1979. *Sophie's Choice.* New York: Random House.

Thompson, Allen. 2009. "Comments on Gardiner's 'Should We Arm the Future with Geoengineering?'" Inland Northwest Philosophy Conference, Pullman, WA, April.

Victor, David, M. Granger Morgan, Jay Apt, John Steinbruner, and Katherine Ricke. 2009. The Geoengineering Option: A Last Resort Against Global Warming? *Foreign Affairs* 88 (2): 64–76.

Von Bulow, Michael. 2009. Failure in Copenhagen Is Not an Option. COP-15 website, October 2 (accessed December 8, 2009).

Williams, Bernard. 1976. Moral Luck. Reprinted in *Moral Luck.* Cambridge, UK: Cambridge University Press. Originally appeared in *Proceedings of the Aristotelian Society,* suppl. vol., 50 (1976): 115–135.

Young, Iris Marion. 2010. *Responsibility for Justice.* Oxford, UK: Oxford University Press.

IV

Reorganizing Institutions to Enable Human Virtue

13

The Sixth Mass Extinction Is Caused by Us

Jeremy Bendik-Keymer

Bureaucracies typically separate men and women from the human or economic consequences of their actions.
—Robert Jackall, *Moral Mazes*

Earth thought 52 of 365: A city of open houses. Life passes through, like utopia. And yet this city is real, over a billion years old. The doorway is the species. A species is evolution's doorway for the ongoing stream of life. It is where evolution happens, what lets evolution through. When we make species extinct, we slam doors shut. Mass extinction closes most of the city for millions of years.
—a journal entry while writing this chapter

Mind the Gap

As you leave the platform for the subway in London's Tube, you see a sign at your feet: *Mind the Gap*. In all the societies in which I've lived, there is a gap *between character and consequence*, though we are not minding it. The gap comes to the fore, to me, with the sixth mass extinction, although we can see it in climate change.

(1) Is it reasonable to blame the gap on bad character?

(2) If not, how should one explain the gap?

My answers are:

(1*) Not always. Reasonably decent people can act wantonly.[1]

(2*) The gap between character and consequence merits study by the social and natural sciences.

My argument falls in behind Bernard Williams's call for the social sciences to inform the field of Ethics in *Making Sense of Humanity* (1998). It also opens the door to natural science informing Ethics, as, for instance, in Matt Ridley's *The Origins of Virtue* (1998).

The conclusion at which I arrive in this chapter is:

(*) Virtue needs to be seen in context.

By "context," I mean:
The <u>reasonable possibilities for action</u> within the institutions and bodies we inherit.

For instance, in the sense of "context" I am using, we act within the context of the natural world, where, for instance, atmospheric changes normally occur invisibly over large spatial and long temporal scales. In such a context, it is not possible for an ordinary person to alter, all by herself, the atmosphere (she could, of course, alter her own contribution to greenhouse gas emissions; but that couldn't alter the atmosphere all by itself). Thus, we don't think that someone is vicious (in the technical sense of having a vice) because she didn't try to change the atmosphere all by herself.

The obviousness of this example shows that we commonly conceive of virtue in context. Aristotle (ca. 350 BCE) thought of virtue as an excellence for the natural kind of being we are, in other words, the human being. He would not have thought our virtue lacking if we could not jump more than a dozen times the length of our bodies, as many spiders can. Why, then, does virtue need to be relocated in context?

Sometimes, new moral problems (e.g., climate change) challenge human nature, which I will call "first nature" ("second nature" being our characters). Sometimes, the combination of these same problems and institutional organization (e.g., mass bureaucracy, mass democracy, globalization) make it hard to see virtue clearly. This chapter examines cases where the kind of moral problem we're facing makes it hard for us to get a grip on where virtue begins and ends due to the limits of our first nature and the limits imposed on us—usually without our awareness—by our institutions. Problems like the sixth mass extinction—and its related industrial problem, climate change, which is greatly causing it—throw virtue out of focus, both because they challenge our cognitive capacities and because they come up hard against the way we organize our world. My strategy in this chapter will be to point to where our focus is lost, thereby generating—so I hope—an *aporia*. I'll use the sixth mass extinction as a case study, because I believe it to be both the most morally invisible of mass-scale environmental problems shaped by the tragic conjunction of industrialism with the limits of our human nature, and because I believe that—from the perspective of our characters—it is among the most marring things we humans are doing (Gardiner, chapter 12, this volume).

Seeing the *aporia* I have in mind is crucial for adapting humanity in the coming centuries. Adaptation is not successful without it being morally acceptable. The economistic "adaptation" Thompson and I discuss in our introduction to this volume is not *adaptation* in my understanding: it is selling out morality to greed (cf. Kawall, chapter 11, this volume). At present, we are hurtling a good deal of the Earth's living forms into the abyss (Wilson 1999). This should be morally unacceptable to most of us (Bendik-Keymer 2006, lecture 6; 2010a), but we are doing it with a bang, without so much as a whimper (*pace* Eliot 1925). Why? And what should we do about it? There can be no adaptation without answering these questions.

One answer is to blame the behavior on vice. But if my argument here is sound, that will not do. Vice is not able to account fully for moral problems like climate change—and that means, too, that virtue is similarly inadequate to addressing them. All the king's horses (such as virtue ethics as commonly understood; cf. the introduction to this volume) and all the king's men (such as all the chapters in part III of this volume) will not be able to put morality back together again.

My argument here comes out of profound frustration with the *limits* idealistic people face and with the conclusion that these limits are not simply the results of not trying hard enough or thinking sharply enough. I reject self-hatred, since I acknowledge our profound overdetermination in the structures that shape our form of life, and I accept our tragic limits as an evolved species with constitutional myopia toward the present (Pinet 2009). Just as the demands of character can keep one up at night, so I have found that the *impotence* of character and the darkness surrounding its constantly undermined intentions can do so, too. I believe it is bad character not to acknowledge that impotence.

The Sixth Mass Extinction Is Caused by Us

What is the sixth mass extinction? As Niles Eldredge (2001) explains:

There is little doubt left in the minds of professional biologists that Earth is currently faced with a mounting loss of species that threatens to rival the five great mass extinctions of the geological past. As long ago as 1993, Harvard biologist E. O. Wilson estimated that Earth is currently losing something on the order of 30,000 species per year—which breaks down to the even more daunting statistic of some three species per hour. Some biologists have begun to feel that this biodiversity crisis—this "Sixth Extinction"—is even more severe, and more imminent, than Wilson had supposed.[2]

Over the past decade, increasing attention has been given to this "extinction event," because climate change will hasten it immensely (Walsh 2009).[3] Recently, the Royal Society B claimed that "the current rate of species extinctions far exceeds anything in the fossil record" (Magurran and Dornelas 2010), and this led climate commentator Joseph Romm to conclude that "the mass extinction we are causing will directly harm our children and grandchildren as much as sea level rise" (Romm 2010). In Romm's evaluation, "Given the irreversibility of mass extinction, and the multiple unintended consequences it engenders, it must be considered one of the most serious of the many catastrophic impacts we face if we don't act soon" (Ibid.). There are more moderate views than Romm's, including criticisms leveled at the science of extinction measurement, but there is no fundamental disagreement that extinction rates are significantly higher than they would have been if humans were not on Earth. The arrival of the sixth mass extinction is a settled belief despite disagreement over particulars.

Like climate change, the extinction is anthropogenic. We have caused it over thousands of years through such things as hunting, farming (monoculture), population growth, habitat destruction, pollution, industrial fishing, and, above all, climate change (Eldredge 2001; Weisman 2007; Pinet 2009; Museum of the Earth 2009). Moreover, climate change is—and has in the past been—a main driver of extinction, with climate-related species loss of 20 percent and upward estimated under current global warming forecasts for this century (IPCC 2007; but this estimate is low). Visualizing the sixth mass extinction is not the purpose of this chapter, but should be done to appreciate its enormity (cf. Bendik-Keymer 2010a). My task here is to understand what our massive extinction of Earth life shows us about the limits of virtue.

Our Acting Wantonly Tends to Destroy Our World of Life

We humans have caused and still cause the sixth mass extinction. This behavior is wanton. Wanton behavior is behavior that displays wantonness.[4] In this chapter I will show that reasonably decent people can nonetheless exhibit wanton behavior. Such a situation one might call "structural wantonness" to disambiguate it from the idea of a character vice. It is structural, because it is not a result of intentions, or even of their failure, but of the (institutional or biological) context in which reasonable possibilities for action appear.

Let me begin, however, with wantonness as a vice. When people speak of wanton behavior, it is usually of "wanton destruction." Wanton behavior is a kind of thoughtlessness, as when we exclaim, "That was such a *thoughtless* thing to do." Wanton behavior is given to destruction by the way its thoughtlessness loses touch with what matters in our world and so tends to damage it.

Here is an ordinary language definition of wanton behavior:

wanton, adjective. 1. *wanton destruction*. deliberate, willful, malicious, spiteful, wicked, cruel; gratuitous, unprovoked, motiveless, arbitrary, groundless, unjustifiable, needless, unnecessary, uncalled for, senseless, pointless, purposeless, meaningless, empty, random; capricious. *antonym* justifiable. (Oxford 2005–2007)

What I want to take from these definitions is the idea of arbitrariness, not the ideas of deliberate destruction. In fact, one of the puzzles of this chapter is *how* people can produce destructive consequences without deliberately doing so. Wanton destruction is arbitrary in the sense that no ground, justification, need, necessity, call, sense, point, purpose, meaning, substance, or cause is said to give it reason. The idea here is that someone has acted arbitrarily without a justification in sight.

We should depart from the ordinary language definition also in the kind of reason that is missing when one behaves wantonly. The ordinary language definition equivocates between whether wantonness is "unjustifiable" and whether it is simply unjustified ("groundless"). It isn't clear whether wantonness involves a failure of moral knowledge or a failure of integrity. However, just because someone lacks a good justification for her behavior doesn't make her behavior wanton. It is when she doesn't have even *her* justification for her behavior that she is (Williams 1981).

Here is why. First, how is simply doing something unjustifiable behaving wantonly? Highly thoughtful and deliberate people do unjustifiable things all the time, not least due to ignorance. Leopold shot the wolf, thinking it would be good for a range of human–environment interactions (see the discussion in Hirsch and Norton, chapter 16, this volume). But he didn't have a frame of mind that allowed him to anticipate the cascade effects of the culling he advocated. He acted ignorantly, not wantonly. By contrast, he would have acted wantonly if he had had the apt metaphor for working with mountain ecosystems, yet put them out of mind.

Second, wantonness is an internal relation between what we think we have reason to heed (including our values) and how we act. It is like weakness of will—*akrasia* (Aristotle ca. 250 BCE). Behaving wantonly is a particular way of being out of control (Frankfurt [1971] 1988). It differs from *akrasia*, in that one need not fail one's better judgment out of emotional imbalance (although one could), but can do so (also) as a result of other sources such as the ones this chapter will later propose. Furthermore, some of those sources may disqualify the behavior as a character flaw, which *akrasia* is. And one is aware of one's better judgment in *akratic* action, whereas in wanton action, it is out of mind.

Wanton behavior is done "without a thought." It is not that nothing goes on in the brain (clearly, it does when we act), and it is not that the act is involuntary. Rather, no attempt at justification is given. Call such an attempt at justification a "rationalization." Thoughtless action, then, is unrationalized. A person is "thoughtless" whenever he acts without having or considering what is—to him—a good enough reason to act. A person is thoughtful *if she has considered what she takes to be the morally salient factors in a situation and has at some point made a good faith effort to justify her action in light of them.* This is a weak definition of thoughtlessness because it is framed subjectively—but the weak version is all one needs to see the gap.

Suppose you are a single mother who has to go to work tomorrow, and your four-year-old son comes down with a fever. His fevers tend to spike high, and so you have to be careful for him. As it happens, H1N1— swine flu—is also in season, and your school has a twenty-four-hour no-fever policy. Suppose, then, you do either of two things: you do not make a point of getting up in the night to check his temperature every few hours, or you send him to school anyway once his fever is gone. You do not really think about the danger he or others might face as a result of your (in)action. That is thoughtless.

Suppose, instead, that before your child goes to bed you give him medicine that has always worked in the past. Waking up in the middle of the night to check on him only once might be careless, but it is not thoughtless. Similarly, suppose you know the course of his fevers so well that you figure this time he doesn't have H1N1. As a single parent, you are at risk of losing valuable income by missing work to take care of him. So you send him to school. Again, this behavior might be careless, even wrong (did you call friends who might watch him in a pinch?), but it isn't thoughtless. You made a good faith effort at being reasonable.

Because wanton behavior is thoughtless, it is insensitive to what is salient in our world. The salient appears in our considerations: "My child is sick." "H1N1 is in season." "Valuable income could be lost." And so on. Each serves our reasoning by tying our world to our goals: "Because my child is sick, I need to get up in the middle of the night." When thoughtlessness ignores moral reasoning, then, it also ignores what matters to us in our world, so long as what matters in our world is not already—easily—made a part of our habits (which do not require reasoning to track reasons). But the problems wherein the gap appears are not tracked in our habits and so demand reasoning. Accordingly, thoughtless people end up not considering what they would think, on reflection, matters.

Because wantonness is insensitive to what we, on reflection, take to matter, it is also linked to destruction. All insensitivity to our world's order isn't destructive, but if we are insensitive to our world when we act, we risk being destructive. Without minding things that matter, we risk damaging them, as when someone puts his child's match-stick collage on a pile of papers that gradually gets buried under more papers until the collage is crushed by a book. "Wanton destruction" is a common use of the adjective "wanton," because wanton behavior leads to destructiveness over time. Wanton behavior runs roughshod over the world.

Our world involves other living things: it is a "world of life." Our world of life reflects our moral stance toward life, whether and where we take living things to be morally considerable. Yet since wanton behavior is given to ignore the order of our world, it is especially dangerous to our world of life. Our acting wantonly tends to destroy parts of it.

This last point is important, because it lets us see the gap and move from wanton character to structural wantonness.

The Gap Appears Only If We Think that Life Is Morally Considerable

We've just considered wantonness as a form of behavior that could display vice. However, when considering the sixth mass extinction, it appears that while our behavior is wanton, our character needn't be. How can this be?

It can be if:

1. Our causing the sixth mass extinction can be wanton, rather than ignorant.

2. Our character can, at the same time, be reasonably decent.

Can our behavior causing the sixth mass extinction be wanton, rather than ignorant? When our behavior is wanton, we disregard something we ought to consider. We don't give it so much as a thought and don't rationalize our behavior regarding it. This mental lapse is different from ignorance in that what we disregard is something we recognize, upon reflection as something we ought to consider.

Imagine, then—for the sake of a ready example—that you are a member of one of the world's major religions, which all hold that living things give us reason to treat them as respectfully as we can, subject to justified use (Jenkins 2010). According to your religion then, the Earth's living things are part of our moral order. They give us reason to stop and think about how we affect them.

For affecting them we are. We are wiping out, in this century alone, and only from climate change (let alone population growth, real estate development, monoculture and toxic pollution), 20 percent and upwards of all Earth life forms (IPCC 2007; compare Magurran and Dornelas 2010). And your actions contribute to this consequence, for instance, by consuming as a typical participant in the U.S. economy in the way U.S. citizens typically consume (e.g., carbon-intensively, relying on agribusiness, etc.).

Now imagine that you exist in a public culture in which these facts are true (as they are in the United States): mass extinction effects have been known for over thirty years (Barney 1980, Ehrlich and Ehrlich 1981), and our potential for driving species extinct was popularized earlier by figures such as Rachel Carson (1962). You have an Endangered Species Act (1973). Moreover, the sixth mass extinction is highly visible—on the front cover of your nation's most iconic magazine (Walsh 2009). There is no absence of ready information.

Next, imagine that you have come face to face with one of these sources of information regarding extinction, at least once, and that when you did you had no obvious reason to dismiss the truth of the claims being made in them.

Finally, imagine that you put off thinking hard about the matter, or simply put down the magazine, turn the news report off, or leave the dinner party behind where the news of the day was discussed. . . . So you leave behind thinking about the grave matter that just crossed your mind, although it would be good to have the time to think about it.

All that's needed is a good faith effort at justifying your effect on endangered or extinct life forms. But you don't make one. And so your faith is out of harmony with your behavior.

We've just considered a thought experiment. But I can report personally that some people do not give their extinction-causing behavior so much as a thought even though they have heard of it (or at least some bit of that behavior, such as contributing to overfishing) and do think that life is the kind of thing deserving some respect. When one pushes them on the matter, they concede that they hadn't really taken the time to put two (their beliefs) and two (their behavior) together. They were just too busy, too caught up in the many other things that make a life and its responsibilities.[5]

They were wanton. And so our behavior causing the sixth mass extinction can be wanton, rather than ignorant.

Is She Decent?

But we still have a candidate character vice. Claim (2) needs addressing: can our characters be reasonably decent when we behave wantonly?

I use "decency" differently than do some philosophers (Rawls 1999; Margalit 1996). For them, decent people (or states) won't humiliate others, whereas I think decent people make a good faith effort at being moral, combined with having no grave vice that incapacitates that effort from gaining ground (i.e., from making moral progress). Mine is closer to an ordinary language definition (Oxford 2005–2007, which emphasizes "conformity," whereas I emphasize potential for growth, cf. Bendik-Keymer 2006, chaps. 6 and 7). Of course, then, if we accept that someone is decent by my definition, she will tend to be virtuous.

Everything then hangs on the reasonable. What is a reasonable good faith effort at being moral? Are you reasonably decent if you don't give a damn about anything? No. How about if you don't go the extra mile to be careful with things? The verdict is out, since it depends on whether those things matter. And how much. Someone who didn't go the extra mile for her child wouldn't be reasonably decent. But someone who didn't go the extra mile for a gallon of gas might be. What is reasonable when it comes to decency is not clear-cut.[6]

On the one hand, this is because the effort implied by decency translates into concern, which is a matter of degree and focus of attention. The life forms we're driving toward extinction matter to us. But how much? If they matter more than anything else, then one can't be reasonably decent and behave wantonly. But of course they do not, at least to most humans, if only because taking care of oneself and one's dependents is a precondition for having a wider life with, for example, other living forms. But perhaps that is indecent? To be decent in a busy world, is our concern for endangered species supposed to burn white hot?

On the other hand, what is reasonably decent is not clear-cut, because, classically, virtue is an excellence (Aristotle ca. 350 BCE). It is supposed to be better than mediocre. But then why let people *be* anything other than excellent when it comes to decency? Perhaps we can't be decent if we don't go the extra mile even in a busy, overdetermined world where we might tend to forget about all sorts of important things. In this light, being reasonably decent depends on being excellent at attending to what matters, and that would seem to preclude that one could both do so and be wanton. This is Kawall's position (chapter 11, this volume) and would appear to be Gardiner's (chapter 12, this volume) as well.

The idea of the reasonable functions here tenuously. If we think that it is reasonable to expect excellence or white-hot concern from people when we call them "decent," then the gap will disappear.

But imagine the single mother again, the one with the feverish boy. Both she and her boy think that life is morally considerable. For instance, they love to take walks in the park experiencing the birds and the wild roses, and they would never hurt or damage them. This mother also seems to try hard to do the right thing, even while busting her butt at work: she's raising her son, providing. She takes some—very little—down time to dance and to date. She makes sure the car is fixed, cleans her own apartment. She volunteers at her son's school. She makes sure her retired parents are all right. She canvases for the nation's first black president, because he seems wisest among the candidates.

However, she also flies in airplanes, eats vegetarian meals with produce that is often flown in, and drives along roads that long ago destroyed much of the local ecology. Although she is a feminist, she is not doing anything now to educate women in underdeveloped areas of the world and to thereby slow population growth. Instead, she counsels local women who are infertile—part of her job as a therapist—and helps women reconnect with their bodies. She also meditates, does yoga, and tries to make up for the various limitations that we all pick up as part of our identity formation. Yet though she cares abstractly about it and spent time learning about it in college, does she now much stop to think about species extinction, let alone mass extinction? No.

Is she decent?

Where Is Character Found?

Yes, she is decent. This real-life example discloses the possibility of wantonness *not* (necessarily) *being a character vice*.

I would suffer from a vice—moralism—to judge that the woman is indecent. In Samuel Sheffler's (1992) terms, I would be inhumane—

overdemanding. This overdemanding-ness could take two forms, and either is damning to the moralizer.

First, my demands could legitimate misanthropy. If the single mother and all those like her are vicious—as Kawall (chapter 11, this volume) and Gardiner (chapter 12, this volume) imply—then hating humanity seems reasonable. After all, if someone who tries that hard to do the responsible thing is actually vicious, then in all honesty, isn't humanity damnable? But it is inhumane to be misanthropic –a point with which Gardiner would agree.

Second, my demands could equate the virtuous person with someone who cannot live a dignified human life. If only the radical activists are virtuous—for example, those who free themselves of enough of the demands overdetermining and limiting the single mother's time to go "off the grid" and "into the wild" or to immolate themselves, literally or figuratively, in protest against extinction and public advocacy—then the problem is not that virtue is exclusive (a common criticism against Aristotle) but that virtue precludes having a life worthy of human dignity (Holland, chapter 7, this volume). The virtuous person, then, *cannot exercise basic human capabilities* such as those involved in responsible parenting. This is a reductio ad absurdum of both Kawall's and Gardiner's positions. After all, while Kawall would seem committed to charging the single mother with apathy or sloth—despite her strenuous and stressful juggling of priorities and causes—the main reason for her over-determination is that she is a parent.[7]

As a matter of virtue (for instance, out of humanity, or out of brotherly love), we shouldn't expect people to be saints. <u>My worry about focusing on the "who we are" question</u> (Gardiner, chapter 12, this volume) <u>is that we risk moralism and miss morality's complex relation to its organizational context and to our human nature</u>. Sometimes, the problem is not *who* we are, but *where* and *how* we are.[8]

There is a gap. It poses a problem that should make us lose sleep: reasonably decent character is not enough to prevent us from acting wantonly.[9] Individual decency exists in a context undermining it. Perhaps, then, we should look at the difficulty of living in a bureaucratically organized world whose economic and political systems are highly ill-adapted to the sorts of challenges posed by the sixth mass extinction, and climate change, too, for that matter (Gardiner 2011). And perhaps we should look at the shortcomings of our nature under our second nature (Ridley 1998; Pinet 2009). Perhaps we should emphasize not just who we *are*, but who *we* are, and then who we are in our *nature*.

The gap leads to darkness. Consider four ways to explain wanton behavior:

Fundamentalism   "We're sinners." The source is a condition—sin. We behave wantonly, though we appear decent, because we are sinful. Here, a flaw in our will—for example, disobedience—disconnects our desires from our reasons and makes us prone to do things that go against our sense of what is good. But the problem is that the single mother does follow through on what she thinks is good; the problem is that she doesn't have enough time to satisfy all the demands of a life worthy of human dignity.[10]

Economism   "People are selfish." The source is a motive—selfishness. We behave wantonly, although we appear decent, because we are selfish. Here, a basic motive shapes what we think is our good, such that when we're decent, we needn't give a damn about other forms of life. The problem with this view, though, is that people do give a damn about other forms of life, including the person we just saw who behaved wantonly (cf. also Jamieson, chapter 9, this volume, on the limitations of economism).

Psychologism   "People are narcissistic." The source is a desire—self-love. We behave wantonly, though we appear decent, because our unconscious desires lead us to love ourselves over all else. The problem with this view, though, is that one cannot be responsible in any deep way—for example, as a parent—and be narcissistic. Narcissistic parents, at least when it comes to their kids, are unreasonable. But we were considering a good parent.

Anthropocentrism   "People are anthropocentric." The source is a belief—human-centeredness. We behave wantonly, though we appear decent, because our anthropocentric beliefs lead us to disregard the value of other life forms. The problem with this view, though, is that one can be anthropocentric and still hold that life is morally considerable. And the problem we're faced with is of people who hold that life *is* morally considerable, yet behave wantonly.

Overall, we are confronted with an agent with sound intentions who doesn't suffer from *akrasia*. The problem isn't in her, but *around* her. It's not reasonably possible for her to discharge all her responsibilities—or to keep everything that matters in mind. This leads me to conclude that character—second nature—is not enough.[11]

If Not Vice, Then What?

I believe contributing wantonly to the destruction of much—if not most—of Earth's life forms is a significant behavioral problem, a marring one (Gardiner, this volume). But I am skeptical that it comes largely from bad character.

Character could have a role in wanton behavior. Imagine, a professor of conservation biology who is vain. He might spend his time on his *vitae* and neglect how his actions contribute to the sixth mass extinction. His vanity causes him to be preoccupied with thoughts about advancing himself on occasions when he could be thinking about the consequences of his actions on other life forms. Vice is then a source of wantonness. Unfortunately, this kind of case doesn't close the gap, because we were considering someone as thoughtful as the single mother.

Character could also await a role in wanton behavior. Perhaps we need to reconceptualize character in light of the context disclosed to us by, for example, climate change (see part III of this volume). My concept of what is reasonably decent would be obsolete. The gap I've presented in this chapter might simply be a sign of virtue going out of focus and needing to be refocused by a better sense of our context. I am sympathetic to this possibility, since I believe that the kind of moral problem posed by the sixth mass extinction <u>will</u> require us to reconceptualize aspects of virtue or even entirely new virtues (*pace* Kawall's reading of me, this volume). But the mother is at least as decent and informed as one would expect a good person to be. She struggles not with norms, but with organizational realities. And I wonder, too, whether, there aren't cognitive difficulties in thinking about processes that take thousands of years to unfold. Some people, understandably, see institutions, not individuals, as the place for handling such cognitive challenges.

I conclude that the gap between character and consequence deserves study by the social and natural sciences. At least, as a first step. Explanations may come to an end elsewhere. Consider just some obvious topics of study:

Cognitive limits Does our evolutionary inheritance predispose us to certain forms of myopia or selective attention (Ridley 1998; Pinet 2009)? How capable are we of keeping in view deep time, a global scale, or distant generations? What are the reasonable overload points for the human mind—beyond which more tasks cannot be kept in mind?

Organizational limits   Does the authority structure of bureaucracies make us lose track of responsibility and put problems out of mind, confident that they are addressed when they are not (Thompson 2005)? Does the way we work—such as in habitual multitasking—fragment the self (Jackall 2010)? Does the structure of our institutions make it hard for us to concentrate on the long-term or the large-scale context of our actions?

Political limits   Do some issues remain out of public debate because of limitations in how representative democracy is typically structured? Is the term length of our public officials and representatives in government supportive for addressing long-term problems? Can the interests of nonhumans readily find representation?

Economic limits   Do the ways we structure our economy make it unlikely that we will pay attention to the sixth mass extinction (cf. Jamieson, chapter 9, this volume)? Do we have unregulated tragedies of the commons that contribute substantially to extinction (Vogel, chapter 15, this volume)? Does our ability to externalize costs lead to such tragedies? Do the timetables for economic gains, such as the quarterly return, make it unlikely that we will consider long-term losses (Jackall 2010, 86–89, 95–101)? And does the way we discount the future do so as well (Gardiner 2004)?

Which areas are addressable through character re-formation—and which are not? My argument in this chapter advocates, first, that we study which are which so that we can more accurately support the "character adjustment" section III of this volume advises, and, second, that we square up with the limits to our second nature by addressing structural wantonness. Then, we should see we have also to adjust the *context* of virtue, within which we find our reasonable possibilities for action. We should see that we have to reorganize our institutions to enable human virtue.

There appears to be an institutional level and a natural base that need exploring to explain how reasonably decent people can contribute to very bad things.[12] We need to add a "where we are" and a "how we are" question to the "who we are" one. Things are a lot more indirect than the authors in part III make them seem. <u>Before talk of virtue, we should discern our reasonable possibilities for action in the institutions and bodies we inherit.</u> Anything else is moralism—and impractical.

Acknowledgments

Many thanks to Colgate University's Environmental Studies program, Ned Hettinger, Jason Kawall, The Liberty Fund, Katie McShane, Ayla Qadeer, Holmes Rolston III, and David Schmidtz.

Notes

1. I will make no empirical claims about how many people fit this description. What matters morally and practically is that there are some people who do, for if we do not grant the argument I am putting forward in this chapter, we will risk being moralistic toward them and risk missing an accurate diagnosis of the dynamics leading to mass extinction, climate change, etc. What matters conceptually is that it is possible for people to exist under such a description.

2. As Holmes Rolston pointed out, these figures are largely "back of the envelope" figures, "guestimates." Still, there is substantial agreement among biologists that a mass extinction is underway. See, for instance, Magurran and Dornelas 2010 and Pinet 2009 for recent discussions. For more extended treatment, see Ward 1994. I focus on violence, not *biodiversity*.

3. The best public education on the event I have seen is at the Museum of the Earth (2009), near Ithaca, New York. Philosophy, sadly, lags behind, with the notable exception of Martin Gorke's (2003) work.

4. I use the expression "wantonness" differently than Harry Frankfurt ([1971] 1988) does. He holds an extreme view according to which a "wanton" is not a person. But we can hold many people exhibiting wanton behavior accountable. And many people exhibiting wanton behavior have the capacity to mold their own character.

5. A good place for *x-phi*, experimental philosophy, would be in determining how common such a situation is.

6. This is already a problem for Kawall's account (chapter 11, this volume). It is not fair to charge someone with greed—e.g., its misestimation of value—when there is profound ambiguity about value. Jamieson's account of value uncertainty (this volume) in the face of global environmental problems should temper Kawall's important account.

7. And not of an excessive amount of children. That has also been thought out.

8. Consider this quote from Robert Jackall:

Managers' days are punctuated by quick huddles and endless meetings; the reading and dictation of terse one-page memoranda, devoid of nuance; the scanning of fluctuating market reports; "brainstorming" sessions to surface bright ideas; listening to the presentations of line subordinates about output, productivity, or marketing, or those of staff about suggested guidelines for corporate policies; talking to lawyers about the impact of existing or projected regulation;

evaluating the work of subordinates; buttonholing or being buttonholed by other managers for a word of advice or warning or simply to pass the time of day; and finally, constant phone calls. *Within such a context* . . . issues do not "come at" managers in any integrated, coherent way, but rather in piecemeal fashion. (2010, 88, emphasis mine)

This example paints a picture of life strikingly like the single mother's. It appears that the organizational form of this society has become bunched up across the board. But is this the society of the "globally wealthy" (Kawall, this volume)? Not exactly—it is found permeating global society even in areas of "uneven development" (Castells 2000), such as in the daily life of Dhaka, Bangladesh, where I have (briefly) taught. If there is a *societal-level* problem of greed here, it is in (the knock-on effects of) globalization's productive form. But that demands further investigation (cf. Postone 1996), and it is in no simple way analyzable *in terms of* greedy individuals (surviving on less than $2 a day!).

9. A consequentialist might say that all the gap shows is that the normative theory implicit in the decent mother's life isn't sufficient for being ethical. But her difficulty isn't a failure to think that consequences matter. As with any decent person, she does think they matter (contra Hallie 2001). The problem is that she fails what is of consequence *out of thoughtlessness* rather than out of not having the—purportedly—correct moral theory. Norms are not at issue. How to keep everything that matters in view is.

10. It is possible that a more rational concept of sin might be a proxy for the kinds of error to which our human nature makes us vulnerable. But then we would turn to natural science—to (the Latin) *ratio*—to understand this dogma of faith—of (the Latin) *fides*. My target is fundamentalism, because it holds an irrational concept of sin.

11. One criticism of my view might claim that it simply shows we have not yet arrived at reformulated virtues in line with our nature. Thompson (chapter 10, this volume) would appear to have to support this criticism. Then, realizing our constitutional myopia as a fact of evolved human nature would support the basic idea of part III of this volume: the need to adjust character. But character is second nature. My worry is that our *first* nature may be inadequate to deal with the sixth mass extinction, climate change, and so on. As a result, some form of augmentation, deliberate limiting, or channeling of our nature is called for in our institutions, not simply a changed form of will. Cf. Hirsch and Norton, chapter 16, this volume.

12. Part of what makes me think the gap is real is the combined awareness in Gardiner's work of both the institutional and the individual. Yet he seems committed to moralism as his argument stands at present. I worry, too, that this leads to a misanthropic tone in his work, which would be relieved if he acknowledged the persistence of moral individualism in his otherwise highly sophisticated institutional analysis. Only then, I believe, would he understand properly the tragic sense of our lives in climate change, a tragedy I have tried to underline in the case of the single mother who does give a damn, but who doesn't have the mental or normative space to do everything that she needs to do.

References

Aristotle. ca. 350 BCE. Ἠθικὰ Νικομάχεια.

Barney, Gerold O. 1980. *The Global 2000 Report to the President*. Washington, DC: U.S. Government Printing Office.

Bendik-Keymer, Jeremy. 2006. *The Ecological Life: Discovering Citizenship and a Sense of Humanity*. Lanham, MD: Rowman & Littlefield.

Bendik-Keymer, Jeremy. 2010a. Species Extinction and the Vice of Thoughtlessness: The Importance of Spiritual Exercises for Learning Virtue. *Journal of Agricultural & Environmental Ethics* 23: 61–83.

Bendik-Keymer, Jeremy. 2010b. *365 Earth Thoughts*. http://365earththoughts.blogspot.com (accessed January 8, 2011).

Carson, Rachel. 1962. *Silent Spring*. New York: Houghton Mifflin.

Castells, Manuel. 2000. *The Rise of the Network Society. The Information Age*. 2nd ed. vol. 1. New York: Blackwell.

Ehrlich, Paul, and Anne Ehrlich. 1981. *Extinction: The Causes and Consequences of the Disappearance of Species*. New York: Random House.

Eldredge, Niles. 2001. "The Sixth Extinction." *ActionBioscience* (June). http://www.actionbioscience.org/newfrontiers/eldredge2.html (accessed August 18, 2009).

Eliot, T. S. 1925. The Hollow Men. In *Poems, 1909–1925*. London: Faber & Faber.

Endangered Species Act. 1973. 7 U.S.C. #136, 16 U.S.C., #1531 et seq., ESA.

Frankfurt, Harry. [1971] 1988. Freedom of the Will and the Concept of a Person. *Journal of Philosophy* 68 (1): 5–20. Reprinted in *The Importance of What We Care About*, 11–25. New York: Cambridge University Press.

Gardiner, Stephen. 2004. Ethics and Global Climate Change. *Ethics* 114: 555–600.

Gardiner, Stephen. 2011. *A Perfect Moral Storm: The Ethical Tragedy of Climate Change*. New York: Oxford University Press.

Gorke, Martin. 2003. *The Death of Our Planet's Species*. Trans. by Patricia Nevers. Washington, DC: Island Press.

Hallie, Philip. 2001. Victims in Wonderland. In *In the Eye of the Hurricane: Tales of Good and Evil, Help and Harm*, 93–101. Watertown, CT: Wesleyan University Press.

Intergovernmental Panel on Climate Change (IPCC). 2007. *Fourth Assessment Report*. http://www.ipcc.ch/publications_and_data/publications_ipcc_fourth_assessment_report_synthesis_report.htm (accessed November 30, 2011).

Jackall, Robert. 2010. *Moral Mazes*. 2nd ed. New York: Oxford University Press.

Jenkins, Willis, ed. 2010. *The Spirit of Sustainability: Berkshire Encyclopedia of Sustainability*, vol. 1. Great Barrington, MA: Berkshire Publishing Group.

Keymer, David King. 1977. Character, Motive, and Behavior: Resources of Language, Patterns of Explanation in Three Castilian Chronicles (1454–92). Dissertation. Yale University Department of History, New Haven.

Magurran, Anne, and Maria Dornelas, eds. 2010. Biological Diversity in a Changing World. *Philosophical Transactions of the Royal Society B*, 365 (1558) (November 27): 3593–3597.

Margalit, Avishai. 1996. *The Decent Society*. Cambridge, MA: Harvard University Press.

Museum of the Earth. 2009. http://www.museumoftheearth.org/ (accessed August 24, 2009).

Oxford. 2005–2007. *New Oxford American Dictionary*. Apple software.

Pinet, Paul. 2009. Deep Time: Its Meaning and Moral Implications. *Forum on Public Policy: A Journal of the Oxford Roundtable* (Spring): 1–12.

Postone, Moishe. 1996. *Time, Labor, and Social Domination: A Reinterpretation of Marx's Critical Theory*. New York: Cambridge University Press.

Rawls, John. 1999. *The Law of Peoples*. Cambridge, MA: Harvard University Press.

Ridley, Matt. 1998. *The Origins of Virtue: Human Instincts and the Evolution of Cooperation*. New York: Penguin Books.

Romm, Joseph, ed. 2010. "Royal Society: 'There Are Very Strong Indications That the Current Rate of Species Extinction. . . .'" *Climate Progress*, November 9. http://climateprogress.org/2010/11/09/royal-society-rate-of-species-extinctions -far-exceeds-anything-in-the-fossil-recordo/ (accessed January 8, 2011).

Rousseau, Jean-Jacques. 1762. *Émile, ou de l'Éducation*. La Haye: Jean Néaulme.

Sheffler, Samuel. 1992. *Human Morality*. New York: Oxford University Press.

Thompson, Dennis. 2005. The Problem of Many Hands. In *Restoring Responsibility: Ethics in Government, Business, and Healthcare*, 11–32. New York: Cambridge University Press.

Walsh, Bryan. 2009. "Vanishing Act: How Climate Change Is Causing a New Age of Extinction." *Time* (April 13): 43–50.

Ward, Peter. 1994. *The End of Evolution—On Mass Extinctions and the Preservation of Biodiversity*. New York: Bantam Books.

Weisman, Alan. 2007. The Lost Menagerie. In *The World Without Us*, 53–67. New York: Thomas Dunne.

Williams, Bernard. 1981. "Internal and External Reasons." In *Moral Luck*, 101–113. New York: Cambridge University Press.

Williams, Bernard. 1998. *Making Sense of Humanity*. New York: Cambridge University Press.

Wilson, E. O. 1999. *The Diversity of Life*. New York: W. W. Norton.

14
Human Values and Institutional Responses to Climate Change

Kenneth Shockley

Values and Institutions in a Time of Climate Change

When addressing problems of the scale of those associated with climate change, it may seem that we need to design institutions to help us handle those problems. We need to develop political bodies and social regimes capable of handling what it might seem individual, uncoordinated persons are incapable of handling on their own (Goodin 1992, 1996a). I might personally desire a worldwide decarbonized economy, but unless the institutions supporting the market act appropriately, the actions I take in support of that desire may be impotent. Institutions, particularly dynamic institutions, seem crucial for adapting to our rapidly changing environment. However, even apart from the worrying sense of personal impotence that may seem to underlie this reliance on institutions, such reliance also disturbingly conceives of institutions as entities that act on our behalf. Not only does there seem to be something pernicious or even vicious in offloading our responsibility for resolving the problems we have caused (Thompson, chapter 10, this volume), we might also worry whether institutions are the sort of things that are capable of acting at all. In this chapter, I will explore the problems associated with designing institutions to be collective agents, and point to a more fruitful way of thinking about institutions. I will argue that due to competing pressures on the way we cannot help but design our institutions, they cannot be the agents of change and adaptation we might well wish they could be in light of our present environmental concerns.

In what follows, we shall see that the very reasons that press us toward institutional responses to policy formation and evaluation indicate the weaknesses and limitations of this institutional approach. I shall begin by providing a rough and admittedly abstract sketch of the role institutions play in environmental policy, and then consider certain constraints on the way we "fit" those institutions to the problems they are designed to address. Institutions are, I claim, loci of accountability, adaptability, and stability. I shall argue that each of these three features constitutes a different constraint on institutional design pressure institutions differently, and that the simplest means of resolving the resulting tension is by enabling our institutions to be collective agents (as we can see in the case of the U.S. Supreme Court). However, I argue, there are tremendous costs in taking this approach as a general strategy, and these costs are particularly high when civic engagement or public support is crucial for the success of these institutions. These costs indicate that, generally, this option should be avoided.

What role can or should institutions play in our efforts to adapt to a rapidly changing environment? It seems intuitively clear that in order for humanity to flourish or even survive in an era of massive climate change, we need to rethink both the role of institutions in our response to those changes and the appropriateness of the individual values we express through those institutions (Bendik-Keymer, chapter 13, this volume). After all, our institutions, from local activist groups to the Environmental Protection Agency (EPA) to the United Nations Framework Convention on Climate Change (UNFCCC), reflect not only our social practices but also our individual values, however indirectly. These values may depend on outdated modes of living; we think both too locally, in our inability to see the larger effects of our actions, and too globally, in our inability to see our own contributions to those effects (Jamieson 2008). I will suggest that we conceive of institutions primarily as social instruments for managing ourselves as we adapt to a changing climate, rather than as agents of adaptive management in their own right. The reasons that point us to a new way of thinking about institutions should also encourage us to rethink the values we express in our social practices (Vogel, chapter 15, this volume). I shall conclude by suggesting that a better, more productive role for institutions is as a means of reforming our social practices and individual values. While it may seem that the fundamental problem with our response to climate change (or lack thereof) lies in the policies we enact, reflecting on the interdependence of our values, social practices, and institutions should lead us to wonder whether the problem might be with the values we express through those social practices and institutions.

Institutions and Their Use

It is not an accident that institutions are a pervasive feature of our social landscape. Institutions are systems of informal and formal conventions, practices, and rules.[1] They include not only formal governmental institutions, like the EPA and the Federal Emergency Management Agency (FEMA), but also organized community groups like Sagoff's (2004) Quincy Library Group and local environmental and political interest group organizations. They include large-scale interest-driven organizations like the Sierra Club and even such politically fractious groups as the Earth Liberation Front. In the case of the latter groups, their role as governing or policy-making bodies is undoubtedly limited, but this analysis applies insofar as they operate as bodies that make policy for some jurisdiction—whether governmental or not—or even merely enable the coordination of a set of loosely organized individuals.

There are clearly a number of legitimate reasons for relying on such regulative structures in the process of aggregating the judgments and determinations of individuals. Many social and political circumstances require that our individual decisions be integrated, aggregated, or collectivized. Institutions provide an exceedingly useful means of doing just that. They may serve to gather the judgments of experts (Bocking 2004; Gunderson, Holling, and Light 1995), to ensure democratic legitimacy (Goodin 2007; Dryzek 2005), to maximize stakeholder buy-in (Sagoff 2004; Curtin 1999), or simply as a means of determining the collective decision of the polity (List 2006; List and Pettit 2005). Institutions, including especially but not exclusively formal organizations, provide a means of focusing our efforts and developing policies that reflect those efforts, and they will clearly play a role in our efforts at adapting to our changing environment. Institutions are capable of both adapting to our efforts and providing a means of focusing our individual and collective adaptation, that is, they both adapt and are a means for our own adaptation to the changes around us. Moreover, institutions provide something on which to focus our evaluation of policy.

Institutions are an indispensible tool in the development, implementation, and evaluation of policy. As we struggle to learn from our efforts at adapting to a changing environment it is helpful to think in terms of the institutions that enact or enable policy. How could we evaluate an institution designed to protect wetlands, say, if that institution did not have a track record, a history, of success or failure? Institutions serve as invaluable instruments for social learning more generally. In an era of global environmental change, the need to develop effective institutions is increasingly pressing. They allow for an evaluative focus of our efforts and an expedient means of enacting the fruits of social learning. Through institutions individual decisions, motives, and values can be integrated within a framework that both allows for values to be exemplified, promoted, or expressed on a large scale, and (ideally at least) enables policy to be responsive to a dynamic and changing environment.

As our environment becomes increasingly volatile, our capacity to adapt to that changing environment will be increasingly tested. We need to rethink what it is to manage the environment, and ourselves. Adaptive management encourages experimenting with various strategies to deal with environmental problems (conceived of quite broadly, following Vogel [chapter 15, this volume]) and work toward solutions as best we can. As Norton (2005; also Hirsch and Norton, chapter 16, this volume) characterizes adaptive management, we need to experiment with policy, learn from our mistakes, recognize the importance of differences in place for the success of policy efforts, and situate our local efforts in the context of regional and global circumstances. Insofar as those strategies and the policies used to implement them are successful, we will learn something useful for the generation of future policies; insofar as they are not, presumably (and hopefully) we will learn something else. Following such a program would provide good reason to believe that we are developing policies (as well as practices) that are both responsive to the natural world, and responsive to relevant human interests. Rather than thinking of the formation of policy as the conclusion of deliberation over some matter of environmental concern, adaptive management sees policy formation and implementation as part of a larger process of social learning.

To adapt requires learning from our mistakes. As the environment changes, so must we.[2] As adaptive management relies on experimenting with policy, institutional approaches allow us to have both a common focal point for policy creation and a structure for evaluating the success of our policies. Perhaps more fundamentally, they provide a means of addressing the difficult task of aggregating individual judgments in the context of a wider set of values and our best understanding of the facts on the ground. We need some means of aggregating our individual judgments, and institutions provide a means for doing just that.[3] Institutions therefore appear to be a vital element in our efforts to adapt to a rapidly changing environment. However, the various roles institutions need to play do not always harmonize well.

Fit

The suitability of an institution to its purpose may be usefully referred to in terms of the *fit* of an institution. Some institutions are more suited to certain purposes: one cannot expect an institution to develop policy appropriate for nonrenewable resources, for example, if it is designed to deal with exchangeable commodities. One would not use something as slow and methodical as the U.S. Federal Court system to deal with a sudden environmental crisis like the April 2009 Deepwater Horizon oil spill. One would not use FEMA when the care, effort, and political sophistication of a legislative branch are more appropriate. And one would not want to be subject to the political intricacies endemic to a state legislature when, say, for reasons of health and safety, an independent regulatory agency would be more appropriate. We can think of suitability of an institution to a purpose or task as a matter of *fit*—but fit comes in several potentially competing dimensions (Young 2002, 58; see also Goodin 1996b). We cannot expect institutions to enact or modify functional policy if they are not appropriately suited to their object. If we want to use institutions as foci of accountability and social learning, we need to consider the structural features of institutions indicative of good fit. As the preceding examples indicate, the way an institution is organized should be appropriate to its task. There are three features we should encourage in the formation of any institutional body entrusted with the formation of environmental policy. Institutions should be *stable*, *accountable*, and *adaptable*.[4]

Institutions should be *stable*, that is, they should be adequately coherent. If they are not, they will not be capable of responding in a diachronically coherent fashion (that is, coherent over time) to whatever object they were designed to address. We need something capable of persisting adequately for social learning to take place. For institutional approaches to environmental problems, we need that institution to persist adequately to determine whether it, in some sense, worked—that is, it must persist as the same institution long enough to be held accountable if we are to learn and adapt. Insofar as an institutional regime is in place for the restoration of wetlands in a particular region, say, we need something against which to measure successes and failures if we are to learn how better to restore wetlands. We need something we can judge as having succeeded or failed. We can further see the need for stability in that institutions are generally designed to provide a response, over time, to persistent problems. Yet stability does not require constancy or a lack of responsiveness to changes in the object, but rather a sufficient degree of diachronic coherence with the reasons and justifications used for previous policy decisions. This coherence is needed for accountability, adaptability, and for all the constraints institutions can provide on political power, self-interest, and moral vices (including that pervasive vice Kawall (present volume) calls "modest greed"). Indeed, stability serves to underpin most other features desirable in institutions.[5] Without some form of stability, there can be little sense made of accountability. Without some form of stability, it is difficult to see how there could be something coherent enough even to be capable of adaptation. In Robert Goodin's (1996b, 40) terms, institutions should be "robust" enough to persist despite a range of external pressures.

As noted earlier, institutions need to be *accountable* for what they enable, for what they bring about, or, for those that are able, for what they do in accordance with their design or purpose. Without considerations of accountability, it is difficult even to recognize successes and failures of policy implementation and other institutional activities. Accountability is needed for social learning. The failure of FEMA to respond adequately to Hurricane Katrina and the subsequent attempts to hold FEMA accountable provides an illustrative case (The Economist 2005). We should understand institutions to have a certain causal efficacy, even if they are not full-blooded authors of the events that they, in some limited sense, bring about (Shockley 2007). Some form of accountability, some means of taking stock of our coordinated efforts and attempts to resolve problems, constitutes a central motivation for taking an institutional approach to generation or evaluation of policy, environmental or otherwise, in the first place (Goodin 1996b). While fit involves being appropriately designed or suited for the problem, it also involves being appropriately responsive to the individuals subject to the institution as well as those who give that institution life (see Goodin 2007; Heath 2006). And it involves being responsive to the dynamic object of policy, that is, to whatever the policy is meant to address. As the problems they are meant to address change, sometimes rapidly, most institutions need to be adaptable if they are to serve their function for any length of time; they need to be able to modify themselves or be modified such that they are able to promote their purposes or satisfy the requirements that they exemplify in light of changing circumstances.[6] Responsiveness to dynamic problem sets and changing constituency constitutes *adaptability*.

These three qualities, stability, accountability, and adaptability, are all vital for institutions if they are to serve as enduring instruments of policy formation, maintenance, and evaluation. When we consider efforts to address climate change, an issue where there are many unknowns, it would be unconscionable not to design institutions that are able to adapt. Further, as we try to learn as a species how best to respond to the changes we have caused, our institutions should be accountable; we should design our institutions such that we might be able to learn from our mistakes and failures in the policies we implement and change through those institutions. And neither of these features would be possible without some degree of stability in our institutions.

However, we will see that these criteria generate dangerously competing pressures for institutional design (see also Young 2002, 14). While institutions seem to serve an invaluable role in adaptive management, and adaptation more generally, the different features that make them so useful press against one another. It turns out that there are certain inherent limits to the use of institutions as tools, at least as they are generally conceived.[7]

Tensions and Resolutions

Accountability, adaptability, and stability require appealing to different structural features of institutions. Accountability requires appeal to the design or function of an institution; it pressures institutions toward a form of teleological integrity. Accountability requires appeal to *purpose*. Holding an institution accountable for, say, the policies made through it pressures an institution to match purpose to outcome, to be accountable for what outcome it enables given its design. Accountability seems necessary for social learning, and therefore for adaptive management.

Further, adaptability requires that an institution respond to external features, to the circumstances in which it is situated including the problem or problems it is designed to address (Vogel, this volume; Hirsch and Norton, this volume). An institution that is incapable of being responsive to changes in a dynamic environment will be of extremely limited use. Adaptive management, and social learning, require us to modify our institutions in accordance with their object, that is, whatever problem or situation they are being used to resolve or control. Adaptability constitutes an *external* pressure.

Finally, stability, understood in terms of being consistently responsive to policy, regulatory efforts, or other actions attributed to an institution constitute an *internal* pressure. Institutions should be responsive to previous decisions and policies attributed to them (fallibly, of course). Indeed, without some degree of what we should think of as diachronic coherence it would be exceedingly difficult to say that there is anything that persists adequately to constitute an institution at all. Stability pressures an institution to maintain coherence over time, and so provides some pressure to minimize change. These three features, accountability, adaptability, and stability, are difficult to satisfy simultaneously.

Consider how we might judge the success (or failure) of a climate change regime like the UNFCCC. We might judge such an institution successful if it brought about or generated policy adequate to the ends for which it was designed (of course, there might be other unintended benefits of such an institution). Presumably, we judge the UNFCCC a success or failure in accordance with whether it led to policy that succeeded in limiting greenhouse gas production in some fashion, over some extended time frame. This is a matter of accountability, of holding the institution to its purpose. Yet we might also consider how well the institution responded to changing pressures from those subject to it—various pressures, including demands (legitimate or not) of fairness and equal consideration and the like—and to the dynamic feature of the environment that it is designed (or merely functions) to address. Along these lines, we might ask whether the UNFCCC can be modified over time to address changing political and environmental conditions. This question involves matters of adaptability, and the pressures on an institution to respond to external considerations. Finally, we might worry about how well the institution has maintained or persisted over time in light of the various pressures on it. We might judge an institution, for example, according to its capacity to survive the very real threat presented by those who resist it or conspire to undermine its functionality. With respect to this dimension of fit, we might ask whether the UNFCCC can stand up against significant opposition. Is it too fragile an institution to survive such opposition? For an institution to be of lasting worth, it must endure and have some capacity to endure in the face of competing pressures (of course, not all *should* last). This constitutes an internal pressure. Most institutions will face pressures from purpose, from external factors, and from internal constraint. These will clearly be in tension.

One way these three conditions might be satisfied simultaneously would be to compel agency on the relevant institutions. We press certain institutions and organizations, like the U.S. Supreme Court and the EPA, for example, toward agency. Just as individual agents satisfy the conditions of adaptability, accountability, and stability, then so too would collective agents (Pettit 2007; Gilbert 2000; May 1990, 1992; Bratman 1999; Tuomela 1995, 2007). If an institution can satisfy the conditions of collective agency, then it should be capable of being accountable for what it does, adaptive to features of its surroundings and to changes in its constituent members, and stable enough to be coherent in its reason responsiveness (that is, acting and responding in a manner consistent with their previous decisions over time). In the least, collective agency seems an ideal toward which institutions should be pressed if they are to be centers of adaptive management and social learning.

However, compelling collective agency on institutions as a general strategy is highly problematic, both in terms of the requirements for institutional agency and in terms of the purposes for which institutions are appropriately designed. It has been shown that an institution satisfying typical requirements for collective agency, and in that way able to adapt to evolving features of the problem it was designed to address, would be decreasingly capable of being responsive to those individuals that enable its collective agency (Pettit 2007; List 2006). This brings us to what I will call Hobbes's Problem. The price of the agency of a collective is some loss of the agency (or at least the autonomy) of those who constitute the collective (Hobbes [1668] 2009). The more the U.S. Supreme Court pronounces decisions as a collective unified agent in its own right, the less the voices of individual justices can be heard in those pronouncements. Of course, those voices are heard in deliberation, and in dissenting opinion, but when the Court decides, the voice is *of* the Court. The actions *it* takes are of the Court, understood as a collective entity. The individual voices are subservient. This loss of individual voice (or autonomy, or even agency) may not be problematic in certain cases as a matter of practice, and may even be a benefit in particular instances (like the Supreme Court). However it does point to a worry. Institutions pressed toward collective agency as a means of being maximally responsive to various external pressures (say, changes in our understanding of the environmental problems those institutions were designed to address) while still maintaining diachronic coherence, as well as some accountability to the original function or purpose, will be able to do so only if those subject to that institution are willing to sacrifice some capacity to express their own autonomy in that institution. They must sacrifice their individual control over the institution of which they are a part. This sacrifice need not require anything as worrying as the complete loss of individual freedom or a collapse of individual will. But it does require that those individuals sacrifice their individual reason responsiveness for the sake of the collective. More specifically, recent work on judgment aggregation (List 2006; Shockley 2009) has shown that either we allow institutions to maintain their coherence at the expense of the coherence of those individuals that constitute them, or those individuals maintain their discursive coherence at the expense of institutional (that is, collective) coherence (List and Pettit 2005; Pettit 2007; Pettit and Schweikard 2006). An institution cannot be fully responsive as a unified entity to the environment while still being fully responsive to the discrete individuals that constitute it. Clearly, this is important for environmental policy-forming bodies, or any bodies dealing with policy needing a degree of public endorsement for success.[8]

It follows that if an institution is capable of balancing the competing pressures of adapting to its problem, of being accountable to its function, and of maintaining stability over time only by means of being attributed collective agency, then it does so at the cost of being responsive to its constituent members. This leads us to the following unfortunate conclusion: the more an institution is appropriately object focused, that is, the more it is responsive to the features of the environment it was designed (or had been allowed to develop) to address, the more it is incapable of being responsive to its members. Even worse, the less it is capable of truly adapting to its environment, the less it is susceptible to pressures of civic engagement (Shockley 2007). This poses a very real challenge. In the case of large-scale problems like climate change, cases involving a large or even global constituency, these pressures become increasingly difficult to accommodate simultaneously. Pressure toward collective agency requires a degree of coherence that seems impractical in such large-scale contexts.[9] Yet social learning seems to require something like accountability and institutional coherence. The cost of collective agency will be unacceptable for many forms of institutions, especially in cases where civic engagement, public participation, and common endorsement are crucial for successful policy. And, of course, this set of conditions is characteristically significant for environmental policy.

Rethinking Institutions

Despite these concerns and the tensions endemic to collective agents, insofar as there is reason to hold our institutions accountable, collective agency might well be an *ideal* toward which we should press our institutions, at least in certain cases. Institutions may often need to be maximally responsive to rapid changes in the environment, and this is best enabled by collectives with the attributes of agents. Further, as we saw in the Supreme Court example, the coherence of an institution may be of such importance that the costs of collective agency are acceptable. But these cases are rare. While the pressures toward collective agency are progressively more difficult to satisfy in global contexts, in just those global contexts we are more in need of some form of regulation as well as some institution to enact and perhaps enforce that regulation (Keohane and Levy 1996). Coherent bodies of policy and coherent policy-making bodies allow for measures of effectiveness and accountability. In such contexts coordination of action, and constraint on action, seem particularly pressing. Institutions are unavoidable features of our response to environmental problems. In certain cases, the sacrifice of individual autonomy *may* be acceptable. However, as environmental concerns become increasingly global we may not only find such a sacrifice unacceptable but also look for institutional structures that are *more*, rather than less, encouraging of civic engagement. Following Sagoff (2004) and Curtin (1999), we should not expect institutions or policies to persist, let alone be successful, without high levels of civic engagement. As reflected in Schlosberg (chapter 8, this volume), public involvement in various forms can be expected to become increasingly important to environmental policy in these transformative times. It is hard to know how to square the need for this sort of engagement with the need for robust institutions.

I suggest we rethink the role played by institutions. If we consider institutions as mechanisms designed to serve as coordination systems, cooperation strategies, and dispute resolution schemes for embedded social practices, then we may have a role for institutions that still allows for social learning, and some degree of adaptability (and even stability), without relying on the dangerous ideal of collective agency. In short, if we think of institutions as tools by which we organize, coordinate, and even cooperate within a larger context of social practices, many of which need to be revised, reformed, or eliminated, then we may have another role for institutions that sidesteps many of the problems associated with institutional agency. We should, I suggest, focus on reforming our social practices, those ways we interact with one another in the process of making social and political decisions. We can think of institutions, on this account, as one key element in the resolution of (or adaptation to) environmental problems. Institutions should be thought of as instruments not only for aggregating individual judgments, but also for refining or refocusing those judgments, for formalizing and making explicit social practices, and, perhaps most importantly, for reforming our social practices and the values that undergird those practices. Why think institutions are suited to this task? One of the basic functions of institutions is to provide a means of constraining ourselves for the sake of some, generally social, good. My suggestion is merely that we conceptualize institutions primarily as tools for reforming our social practices and values.

This suggestion leads us to two general, more practical points. First, accepting this role for institutions requires an increased recognition of the importance of social practices in the development of responsive institutions. Second, shifting the focus from the institutions themselves to values and practices should remind us of a sensible worry about relying on institutional agents to do our work for us (Curtin 1999; Vogel, chapter 15, this volume). This shift is something we should be thankful for: in an age of anthropogenic climate change, it would be exceedingly difficult, as well as politically problematic in the extreme, to create a global institution robust enough to constitute an agent. Of such global collective agents, nightmares are made.[10] Yet more than ever, as we need changes in our social practices as much as changes in our institutional and even political structures, we need institutions to reform, coordinate, and empower our individual judgments and social practices.

Taking Control of Our Institutions, Our Values, and Our Future

We have seen that collective agency comes only at a great cost, and yet often looks to be the best way to design institutions suitable for adaptation to a rapidly changing environment. As instruments of adaptive management, institutions are potent entities; as means of creating policy from our disparate perspectives, they may well be invaluable. Yet while our institutions and collective bodies can, under certain conditions, constitute distinct deliberative entities, we should not use this as a reason to defer control to those collective entities. We need to take control of our institutions, and through our institutions, our social practices and the values those social practices instantiate. We can see the practical significance of taking control of our institutions if we contrast passive preference aggregation (e.g., simply summing the results of a poll over a single question) with deliberative models of decision making (e.g., deliberation by debate in a public forum). If we understand our institutions to be the agents that gather together the simple data we provide (say, through voting), then we have ceded control of that very means of directing our policy formation. If we understand our institutions as means of enabling participants to deliberate about the best norms and policies to guide them, about the values at play in their social practices, then we have a better way of expressing the values about which we care in the policies we form.[11] We should think of our institutions less as constraints on us, and rather as tools we actively use to express and revise our values. Following Vogel's (chapter 15, this volume) themes, we should think of institutions as constructs that bind us to our social context, and, though that context, to one another.

Reconceiving institutions in terms of their role in expressing and reforming values and social practices should encourage us to accept two general conclusions:

1. It is our not our institution's responsibility to bring about the changes we deem necessary. *We cannot offload responsibility without offloading control.* Institutions are merely tools.

2. Social practices should be the focus of our concern. Institutions should provide means of reforming our forms of interaction in light of our shared values and shared concerns. *Institutions should be thought of in terms of their use for self-regulation, not for authoritative compulsion.*

Yet there is a danger of making our institutions *too* passive. If we make them too weak, such that they do no more than follow civic interests and vanish with those interests, they will lose their usefulness. Allowing institutions associated with the formation of viable policy to be this ephemeral could well lead to catastrophic consequences when urgent action is needed, for this is often when we most need to constrain ourselves. Institutions need to motivate action on the basis of real concerns, bringing our attention to our own values and holding our own heels to the fire when those values are inconsistent with our practices. We should be reminded of the values we express: if we purport to value clean air, and our social practices have not pressed us away from polluting practices (say, the practice of purchasing the largest car one can afford), then our institutions should reflect this tension (say, by making socially apparent the contrast between emissions standards and our own individual emissions, especially when aggregated). Institutions should serve to focus us, and to motivate us, through the reformation of our social practices and the virtues and values we honor. We might ask of our social practices whether they are too passive or wasteful or reflective of unhelpful values given current conditions.

Considering the coherence of our values and social practices brings us back to the problem of climate change. We need to ask ourselves whether the values we understand to be constitutive of human flourishing mesh with our practices of, for example, building, transportation, and employment, given what we know about both their contribution to greenhouse gases and the probable effects of climate change. If they do not (and they do not), then our social and political institutions ought to be reformed, or new ones created to revise those practices, to make clear the tensions and press us toward solutions. Just as our institutions of medical care and law pointed to an inconsistency between our social practice of accepting smoking as no more than a personal lifestyle choice and a wide range of health values we purport to express, so we might expect institutions to highlight inconsistencies between our social practices and the values associated with our conception of human flourishing.

I have argued that due to structural pressures on institutions, we cannot expect them to serve as agents for resolving our environmental problems without sacrificing our own decision-making authority. As a consequence, I suggested that we need to rely more centrally on social practices. The importance of social practices, coupled with the need for these institutions to be responsive to a rapidly changing environment, presses us to take control of our institutions and ensure they reflect the values appropriate for the problems and circumstances we face. In order for humanity to flourish, or even survive, in an era of massive climate change, we need to rethink basic strategies of adaptation. Rethinking these strategies requires more than simply redesigning our institutions, it requires rethinking the values we espouse and the social practices in which we participate as much as the institutions we create. In an era of climate change, we need to ensure our institutions reflect our values. We need more than coordination to adapt to a changing climate; we need to adapt both the social practices that shape our interaction with one another and the environment, and the individual values that are expressed through those social practices.

Acknowledgments

The author would like to thank Allen Thompson, Jeremy Bendik-Keymer, James Delaney, and the audience at Clemson University for helpful discussion, comment, and critique on earlier versions of this chapter.

Notes

1. Following Oren Young, we should take institutions to be, roughly, "sets of rules, decision-making procedures, and programs that give rise to recognized practices, assign roles to participants in these practices, and govern interactions among occupants of specific roles" (2002, 30; see also Goodin 1996b, 19–20). Institutions are central to social arenas from markets to political deliberation—merely setting the boundaries and determining modes of discourse will generally involve institutions of some form (Keohane and Raustiala 2008; Dryzek 2005).

2. Along similar lines Sandler (2007; chapter 3, this volume) indicates that as the consequences of global warming are felt, the virtues associated with restoration and recovery will become increasingly salient. These virtues, including qualities like openness and accommodation, are equally appropriate for adaptability.

3. An interesting question is whether institutions actually *make possible* what could count as a judgment. Vogel's "constructionist" position (chaper15, this volume) and Bendik-Keymer's "virtue in context" position (chapter 13, this volume) would imply that they sometimes do.—eds.

4. This set of qualities extends well beyond the usual contrast between "the logic of consequences and the logic of appropriateness" (Young 2002, 17).

5. Note that the stability referenced here is necessary for rule of law (Dworkin 1986).

6. This account clearly understands institutions teleologically, following Seumas Miller's (2001) account of social groups. See also French 1984.

7. These inherent limits are in addition to the practical limits that plague institutional approaches to large-scale management. See Dryzek 2005 and Jamieson, chapter 9, this volume.

8. The need to be responsive to the variant pressures of the public is a recurrent theme not only of those who promote civic engagement as a means of addressing environmental problems (Curtin 1999; Sagoff 2004) and the environmental pragmatists (Norton 1991, 2005; Weston 1996; Light 2002). The importance of being responsive to how the policy is put into practice, and therefore how the public responds to the policy, is characteristic of what has been called the new institutionalism (Young 2002; Goodin 1996a; Knight and Sened 1998), an approach that tries to look at the background social context as much as at formal structural models of institutions.

9. See also Jamieson, chapter 9, this volume, for practical concerns involving institutional approaches in such contexts.

10. Although see Hirsch and Norton, chapter 16, this volume, for a well-developed attempt to resolve some of the more nightmarish elements of large-scale institutions.

11. For quite different purposes, a similar appeal to the need for taking active control of our relation with the environment, and of the reappropriation of our

social practices, can be seen in Vogel, chapter 15, this volume. We need to take control of our institutions if we are to prevent them from alienating us from our environment, and from each other.

References

Bocking, Stephen. 2004. *Nature's Experts: Science, Politics, and the Environment.* New Brunswick, NJ: Rutgers University Press.

Bratman, Michael. 1999. *Faces of Intention.* New York: Cambridge University Press.

Curtin, Deane. 1999. *Chinnagounder's Challenge.* Bloomington: Indiana University Press.

Dryzek, John. 2005. *The Politics of the Earth.* 2nd ed. New York: Oxford University Press.

Dworkin, Ronald. 1986. *Law's Empire.* Cambridge, MA: Harvard University Press.

The Economist. 2005. The Shaming of America. *Economist* 376 (September 10): 11.

French, Peter. 1984. *Collective and Corporate Responsibility.* New York: Columbia University Press.

Gilbert, Margaret. 2000. *Sociality and Responsibility.* New York: Rowman and Littlefield.

Goodin, Robert E. 1992. *Green Political Theory.* Cambridge, MA: Polity Press.

Goodin, Robert E., ed. 1996a. *The Theory of Institutional Design.* New York: Cambridge University Press.

Goodin, Robert E. 1996b. Institutions and Their Design. In *The Theory of Institutional Design,* ed. Robert E. Goodin, 1–53. New York: Cambridge University Press.

Goodin, Robert. 2007. Enfranching All Affected Interests, and Its Alternatives. *Philosophy & Public Affairs* 35: 40–68.

Gunderson, Lance H., C. S. Holling, and Stephen S. Light, eds. 1995. *Barriers and Bridges to the Renewal of Ecosystems and Institutions.* New York: Columbia University Press.

Heath, Joseph. 2006. The Benefits of Cooperation. *Philosophy & Public Affairs* 34 (4): 313–351.

Hobbes, Thomas. [1668] 2009. *Leviathan.* Oxford, UK: Oxford University Press.

Jamieson, Dale. 2008. The Post-Kyoto Climate: A Gloomy Forecast. *Georgetown International Law Review* 20: 537–551.

Keohane, Robert O., and Marc A. Levy. 1996. *Institutions for Environmental Aid.* Cambridge, MA: MIT Press.

Keohane, Robert O., and Kal Raustiala. 2008. Toward a Post-Kyoto Climate Change. Architecture: A Political Analysis. HPICA discussion paper 08-01, Belfer Center for Science and International Affairs, Cambridge, MA.

Knight, Jack, and Itai Sened, eds. 1998. *Explaining Social Institutions*. Ann Arbor: University of Michigan Press.

Light, Andrew. 2002. Contemporary Environmental Ethics: From Metaethics to Public Philosophy. *Metaphilosophy* 33: 426–449.

List, Christian. 2006. The Discursive Dilemma and Public Reason. *Ethics* 116: 362–402.

List, Christian, and Philip Pettit. 2005. On the Many as One: A Reply to Kornhauser and Sager. *Philosophy & Public Affairs* 33 (4): 377–390.

May, Larry. 1990. *The Morality of Groups*. Notre Dame, IN: University of Notre Dame Press.

May, Larry. 1992. *Sharing Responsibility*. Chicago: University of Chicago Press.

Miller, Seumas. 2001. *Social Action*. New York: Cambridge University Press.

Norton, Bryan G. 2005. *Sustainability*. Chicago: University of Chicago Press.

Norton, Bryan G. 1991. *Towards Unity among Environmentalists*. New York: Oxford University Press.

Pettit, Philip. 2007. Responsibility, Incorporated. *Ethics* 117: 171–201.

Pettit, Philip, and David Schweikard. 2006. Joint Action and Group Agency. *Philosophy of the Social Sciences* 36: 18–39.

Sagoff, Mark. 2004. *Price, Principle, and the Environment*. New York: Cambridge University Press.

Shockley, Kenneth. 2007. Programming Collective Control. *Journal of Social Philosophy* 38 (3): 442–455.

Shockley, Kenneth. 2009. Environmental Policy with Integrity: A Lesson for the Discursive Dilemma. *Environmental Values* 18: 177–199.

Tuomela, Raimo. 1995. *The Importance of Us*. Stanford, CA: Stanford University Press.

Tuomela, Raimo. 2007. *The Philosophy of Sociality: The Shared Point of View*. New York: Oxford University Press.

Weston, Anthony. 1996. Beyond Intrinsic Value: Pragmatism in Environmental Ethics. In *Environmental Pragmatism*, ed. Andrew Light and Eric Katz, 285–306. New York: Routledge.

Young, Oren. 2002. *The Institutional Dimensions of Environmental Change*. Cambridge, MA: MIT Press.

15
Alienation and the Commons

Steven Vogel

What does it mean to be alienated from nature? The claim that we suffer today from such alienation is familiar in environmental discourse, but it isn't always clear what that means. Both "nature" and "alienation" are famously difficult terms, first of all, but second (as I'll argue), under certain standard understandings of those terms nature is exactly the sort of thing from which one *cannot* be alienated. In its most common interpretation, it seems to me, the claim that we're alienated from nature doesn't actually make much sense. And yet I do think we are alienated from something *like* nature, and that we need to understand and overcome that alienation. What we're alienated from, I'll argue, is the *environment,* meaning by that word something different from "nature." I will end by suggesting that the alienation stems from certain characteristics of our social system—in particular, from the lack of any social way to justify or even to acknowledge the public consequences of private actions—and that it's related to the kind of collective action problem often called "the problem of the commons." Overcoming alienation, on this account, will be less a matter of developing new virtues than of finding ways to transform the institutional context within which human practices take place.[1]

Most frequently it is claimed that we are alienated from nature because we fail to recognize ourselves as *part* of nature. We are natural beings, dependent on natural forces for our existence, and yet we treat nature as if it were something distinct from us, and as something we could (and should) master. We view nature anthropocentrically, this claim continues, seeing it merely as a sort of raw material at our disposal, and the upshot is that we destroy it with technologies that attempt to reshape the natural world into an artificial one structured for human purposes. Yet such attempts are never successful, because in fact we depend upon nature, and so it takes its revenge on us, as our technologies produce increasingly dire consequences. To overcome our alienation would be to give up anthropocentrism and to reintegrate ourselves *within* the natural order, abandoning the impossible dream of replacing the natural world with one created by humans, and learning instead to live in harmony with nature.

Familiar as it is, such an account suffers from significant conceptual difficulties, not least about what exactly it means by "nature." If human beings are really part of nature, first of all, it isn't clear how their actions could possibly destroy it, or why a human-made world wouldn't still be a natural one. The dams that beavers build are natural: why not the coal-burning power plants that humans build? In one standard sense of the word "nature," the fact that humans and their abilities evolved in accordance with the same biological processes as other species means that all human behaviors are natural, which makes it hard to see how our technological behaviors could be said to alienate us from nature. Yet in a second, and equally standard, sense of the word "nature" refers uniquely to that part of the world that has not been affected by human action, which would seem to make our technology "unnatural " by definition. But how could one be said to be alienated from something from which one has been excluded by definition? And in any case how could this definition be justified given the claim that we are *part* of nature?[2] Further, in what sense has anthropocentrism been avoided here—in this definition that grants to one species, alone among all the others, the special ability to change an object from being natural to being artificial, and thereby to destroy nature? When spiders spin a web or beavers build a dam, no one suggests that nature is thereby destroyed; yet the human species, apparently, is different, more ontologically potent than these others.

Thus this familiar idea of what it means to be alienated from nature, despite its insistence that we are part of nature, actually seems to require the assumption that we are outside of it, in which case it isn't at all obvious how our asserted alienation from it could ever possibly be eliminated. Or might there be a third definition of "nature" at work here, according to which our actions are neither all natural (as the first definition would suggest) nor all unnatural (as the second definition would), but instead *some* of our actions are natural and some not? This would seem to allow the possibility that we could overcome our alienation from nature by limiting ourselves to the first kind, resolving to act only in "accordance with nature." But what would *this* definition of "nature" be, and how would it distinguish "natural" actions from unnatural ones?

And how did (natural) evolution somehow produce creatures capable of both natural and unnatural acts?

Note in particular that while it might seem attractive to say that although humans themselves are part of nature, their products aren't—they're "artificial"—a little thought will show that this won't work either. For no one thinks that all the objects human beings produce are artificial: we produce babies, for example, and urine, and carbon dioxide when we exhale, and no one calls these "artificial." It won't do to say that only those objects we produce using "biological processes" are natural, because that requires previously deciding what is to count as a biological process or not, precisely the question under discussion: to say that spiders weaving webs is a biological process while humans building nuclear power plants is not is already to have decided that certain human actions aren't natural. And it won't do, either, to say that only those objects we produce intentionally are unnatural, because after all sometimes people intentionally produce babies (and even, under certain conditions, urine and exhaled CO_2).[3]

The real intuition behind the idea that (some of) our products aren't natural, it seems to me, is a Cartesian one: it's that we are *not* fully part of nature, but rather are some sort of strange inwardly divided creatures that are partly natural and partly unnatural. But how did that happen, in the course of (natural) evolution? And, again, why shouldn't this view, with its obvious debt to a dualism that nowadays hardly any serious philosopher would dare to defend, be called anthropocentric? We certainly are an amazing and absolutely unique species, according to this view: rooted in nature but also somehow possessing "unnatural" parts, and remarkably capable as well of using those parts to transform ordinary natural objects into "artificial" things that are outside nature.

The idea that we are "alienated from nature," I am suggesting, seems to depend on presupposing what is an essentially *metaphysical* distinction between humans and nature. But to be alienated from something isn't only to be separated from it, it is to be *illegitimately* separated from it. And if the distinction between humans and nature is a metaphysical one, then the separation isn't illegitimate but rather metaphysically necessary, which renders it unclear why it deserves the name of alienation at all, or why it should be a subject of critique.

Could we imagine a conception of alienation from nature (or something like it) that did not presuppose such a metaphysical distinction between humans and nature? I want to propose that such a conception can be found in work of the young Marx, and to argue that in fact that conception would be considerably more helpful in thinking about environmental questions, and their relations to political and social ones, than the familiar one I just outlined. Marx's account, as I reconstruct it, has its philosophical roots in Kant's Copernican Revolution, and particularly in Kant's radical suggestion that the problem of knowledge can only be solved if the knower is understood not as passively receiving information about the world (as empiricism does) but rather as actively imposing structuring categories upon the material of sensation. The world we perceive, for Kant (which is to say, the world we *inhabit*), is a world that we knowers have actively prestructured; in that sense it is not something external to us or other than us, but rather something in whose production we are ourselves implicated. Post-Kantians, of course, take this idea and develop it further—most importantly in Hegel, who interprets *Geist's* relation to the world it confronts on the model of this sort of active shaping, and who furthermore introduces the idea that *Geist* could find itself alienated from that world by failing to recognize itself in it (Hegel 1977, 294–295).

Marx's materialist turn can be understood as replacing the obscure Kantian notion of a disembodied knower magically "constituting" the world (or the even more obscure Hegelian notion of *Geist* doing so, and coming to know itself in so doing) by the idea of concrete human beings as active creatures transforming their world through work. We come to know the world by *acting in it*, and to act in it is to change it. Marx takes the Kantian idea of the phenomenal world of experience as something the knower constitutes and reinterprets it as meaning that the practical world humans inhabit is something that they actively *build* (Lukács 1971). And alienation occurs when we fail to recognize the objects we have built as our own products and so they come to appear as external to us, as alien powers over and against us.

Thus Marx reinterprets a series of epistemological categories in terms of *practical human activity*. Knowledge is understood as practice, and more specifically as labor. Accordingly alienation is reinterpreted as an economic phenomenon: it is the worker who is alienated, in that he builds an object through his labor but is unable to recognize himself in it as it becomes the property of someone else, the capitalist, and indeed becomes the basis of the capitalist's wealth, thus turning into the very condition that creates the worker's own poverty. Dynamic human activity turns into an object, or rather into a world of objects—the world of capitalist wealth from which the worker is estranged. Alienation arises, Marx writes, when the object which labor produces—labor's product—confronts it as an *alien being*, as a *power independent* of the producer. The product of labor is labor which has been embodied in an object, which has become material: it is the *objectification* of labor. The objectification of labor is its realization. But under contemporary economic conditions this realization of labor appears as the *de-realization* of the laborer, objectification appears as *loss of the object* and *bondage to it*, appropriation as *estrangement, as alienation.* . . . The *alienation* of the laborer in his product means not only that his labor becomes an object, something *external*, but that it exists *beyond him*, independently, as something alien to him, as a self-subsistent power over and against him. It means that the life which he has conferred on the object confronts him as something hostile and alien.[4] (Marx 1975, 272)

The subject who fails to recognize itself in its product and is therefore alienated, furthermore, is a *social* subject; production, Marx says, is always a collective process. His critique of capitalism's approach to the division of labor centers precisely on this point: although all labor is implicitly social and cooperative, under capitalism it only appears in an individual and private form. Our social relations with each other appear as external to us, as a set of relations (Marx famously writes) not between people but between things. This is the point of the account of value in *Capital*: the "value" of a commodity, Marx argues, is an expression of its social character as having been produced through cooperative labor for use by community members, but it appears in the form of what seems like a natural fact about the commodity, its price (Marx 1977, 164–165). The unemployment rate, the trade surplus, the performance of the NASDAQ: these are all consequences of the implicitly cooperative practices through which we satisfy each other's needs, but they appear instead as a given and unchangeable external context that constrains our practices. They appear, that is, like *facts of nature*, not as the results of our own doings. What Smith called "the invisible hand" is what Marx sees as alienation: when we are only able to act as private individuals, the social effects of our actions appear to us as external facts of nature that we cannot control but to which we must simply adjust ourselves. The "invisible hand" is what results when social action is not recognized and determined as such and so turns into the form of a *thing*—which is to say, turns into something like nature. Overcoming this alienation would take the form of a recognition and reappropriation of these social processes as social, which to Marx means putting them under the control of democratically organized planning processes.[5]

The implications of Marx's account for the question of alienation from nature, and for environmental philosophy in general, are striking.

The moral of his account is that alienation arises when we fail to see the *constructed* character of the objects and institutions that surround us, a constructed character that is inseparable from their social character. We are alienated from objects that we have produced through our own actions, and our alienation arises when we fail to recognize them as such. And when such a failure of recognition occurs, the result is that the objects and institutions look like natural ones. For Marx, then, *the appearance of nature is itself a symptom of alienation.* Alienation occurs when something that is really social, and socially constructed, appears to be natural.

Now at first, this doesn't seem to help much if we're trying to understand alienation from nature—after all, nature surely *is* natural (isn't it?), and it *isn't* socially constructed. And yet there is more to say. For Marx's account, and especially its view of labor, helps us to notice how much of the world we actually inhabit consists entirely of objects that have been built by human beings through human labor. The world that "environs" us—our "environment"—isn't "nature" at all: it's a *built world.* We all know this, of course, in a certain sense, but we don't often pay attention to it or think about its implications—and indeed, we usually treat the words "environment" and "nature" as synonymous. A commitment to "environmentalism" is understood as a commitment to the protection of nature, just as "environmental concerns" are concerns about the effects of human action on nature, and "environmental damage" is damage to the world of nature. In talking as though the things that "environ" us are things of nature—instead of the humanly built things by which we are actually surrounded—is it possible that we are making the kind of mistake that Marx diagnoses? Could it be that we are alienated, not from nature, but from the built world, the one that actually environs us, precisely in that we fail to see it as a human, and more specifically a social, product? Is it possible that a symptom of our alienation is that we think of our environment as *nature*—which is to say, as something that we did not build, and for which we are thus not responsible?

We're back, of course, to the question of what "nature" means. It's true, no doubt, that if it means that part of the world that is separate from humans, then by definition nature cannot be a human product, and so in Marx's sense we cannot be alienated from it. Yet as we have seen, construing nature in this way causes some serious conceptual difficulties, not least because it seems to mean that humans aren't natural. And in any case such a nature turns out to be remarkably hard to find. As Bill

McKibben (1989) argued two decades ago, in the context of phenomena like global warming every spot on earth has already been changed by human action, and so nature can be said already to have ended.[6] Yet even if there's no nature left, there's surely still a world that environs us, an *environment,* and it does not seem so strange to suggest that *that* world is a human product: indeed, saying this was precisely McKibben's point. And so while perhaps given Marx's conception one cannot be alienated from *nature,* it still makes sense to talk of being alienated from the *environment.* And isn't it the environment, after all, that is supposed to be the object of environmental concern?

We have built the environment. (Look around you, reader! Which of the objects environing you was not produced through human labor?) To say this is not to say we have done so ex nihilo, of course: no building is ever ex nihilo. But it *is* a human product: the world that surrounds us has the shape it has and consists of the objects it does because of our activities, our desires, our institutions, our social and economic structures.[7] Yet if the world that environs us has come to be what it is through human action, it is surely also the case that what it has come to be is something pretty bad—dangerously warming, not to speak of ugly, toxic, and harmful to many of the creatures that inhabit it. It is our actions that have led to those results, and yet of course no one ever really intended such results. They are the consequences of our actions, but they seem to us like—well, like "facts of nature." But this is just where Marx's account might be helpful, precisely because it focuses on the question of how the things we produce can come to seem like independent and threatening powers opposed to us. We are alienated from our environment when we fail to recognize it as the product of our own actions and thus fail to acknowledge our own responsibility for it, and so instead it starts to look like a natural fact about which there is nothing we can do: global warming simply part of a natural cycle, pollution an inevitable byproduct of technology, urban sprawl the inexorable consequence of market forces, and so on. It's when the environment comes to look like "nature," that is, that our alienation from it occurs.

But if this is right, then the familiar conception of "alienation from nature" I began by describing—which identifies the environment with nature and understands the latter as distinct from human beings and human practices, and thus precisely as something we have *not* produced—is *itself a symptom of alienation.* By treating nature as separate from the human and criticizing humans for encroaching upon it, such a view sees the world humans inhabit (their environment) in the same way that Marx says that alienated workers see the capitalist social order—as an externally given reality they cannot change but must simply accept, instead of as something that their own practices have helped to produce.

Nature is not distinct from the human, and the human position within it is not one that would make it possible somehow to "leave nature alone" or "let it be." The nature/human dualism presupposed by such locutions makes no sense, not merely because we humans are natural organisms but more importantly because (like *all* natural organisms) our role in the world is fundamentally active and transformative, and so the "nature" we inhabit (which is to say, our environment) is one we have *always already helped to form*. The familiar view of alienation from nature I am rejecting sees alienation as the inevitable result of human action, implicitly suggesting that whenever we transform the world we alienate ourselves from it. But all natural organisms transform the world: to be alive is to act, and to act is to change something. We cannot avoid transforming the world we inhabit, which means that the world we inhabit is of necessity a world that we have always already helped to produce.[8]

Marx's account can be understood as suggesting that the environment we inhabit is "socially constructed." But social construction here doesn't mean, as it frequently does in debates among environmentalists, that our *ideas* or *experiences* of "nature" are socially variable (Soper 1995). Rather here it is meant *literally*: the environment we inhabit is literally built, by the socially organized labor of human beings. To say this is in no way to deny the reality or materiality of the world, or to see us as its masters. Construction—in the physical, literal sense—takes place in a real material world; it's hard work, requires effort, sometimes fails, and always turns out differently than those who engage in it expect. To say the environment is socially constructed is not to say that humans can make the environment any way they want, or to confuse what a society believes about the environment with what it really is. The environment we inhabit is indeed the product of our practices, but this does not mean it is the product of our desires or our beliefs.[9] What we expect or intend to build isn't the same thing as what we actually build: every architect and engineer knows that.

To be human is to transform the world. The view of alienation I am criticizing thinks that we alienate ourselves from nature when we transform it—as if we had any choice in the matter, as if transforming nature were not itself *our nature* (as it is the nature of everything in nature). Alienation, I am suggesting, arises not from our transformation of the world but rather from our failure to recognize ourselves in the world we have transformed—a failure, that is, to acknowledge responsibility for what we have done and built. That means, first of all, not recognizing that the environment that surrounds us—our economic and social institutions, our sprawling suburbs, our rapidly warming atmosphere—is something that results from human practices, and is *not* a "fact of nature." And second, it means not recognizing that the practices that produce that environment are *socially organized* ones. They have social preconditions and social consequences, and operate in accordance with socially defined norms; and further they are almost always engaged in alongside other people.[10] And so our responsibility for the environment is social, not private.

The environment we inhabit is a *built* environment, one which has always already been affected by human action. Our environmental problems arise *there*; they are problems of pollution and global warming and overpopulation and (not least) ugliness and cruelty. They don't arise in a pure "nature" independent of human action. We aren't alienated from "nature" in the abstract (we can't be, really); we're alienated from *that environment*, and not because we don't acknowledge its independence from us but rather because we don't see that it is *not* independent, and so fail to acknowledge our (social) responsibility for it in both the causal and the moral senses of that word. Environmental problems aren't problems having to do with nature, or with our relationship to nature: they have to do with *society*.

We live in a social order where each of us can only act as a private individual, and so the overall social consequences of our actions appear like facts of nature about which there's nothing any individual one of us can do. This key characteristic of what Marx called alienation bears important resemblances to a more familiar idea in environmental discussions—the sort of collective action problem known as the "tragedy of the commons." Garrett Hardin's famous parable of herdsmen whose cattle share a common pasture indeed possesses the same structure: the act that is rational for each herdsman to perform—adding additional cattle to his herd—nonetheless when generalized has consequences for all the herdsmen which they seem unable to avoid (Hardin 1968). I want to conclude by considering the problem of the commons in a bit more detail, and by reflecting on what it means for an account of environmental "virtues."

Hardin's model is clearly relevant to a whole series of environmental problems, and its relevance to the problem of climate change is certainly obvious (Gardiner 2001). An individual factory owner has to decide whether to increase production and therefore emit more pollution into the atmosphere; an individual fisherman has to decide whether to use a new technology that allows significantly more fish to be caught: an individual commuter has to decide whether to drive to work and burn fossil fuels. In each case the benefit of performing the action considered is great, and the cost (in pollution, in loss of fish species, in greenhouse gas emissions) of the individual action is tiny, although it is large in the aggregate. The environmentally concerned industrialist or fisherman or commuter faces the problem of the commons: he or she may genuinely desire to decrease pollution, protect the fish, prevent global warming, but is aware that foregoing the new factory, the new fishing technology, the drive to work, will in fact have *no significant impact* on achieving these goals on the one hand while it will deeply and negatively affect his or her own situation on the other. It would be irrational (and probably economically suicidal) to refuse to perform the action considered, and this would be so *even if every actor had the same commitment to environmental goals*. Left to their own devices, which is to say functioning as private individuals, the actors have no choice but to act in ways that they are fully conscious have disastrous environmental consequences.

I want to emphasize here that the problem does not arise from the selfishness or shortsightedness of the actors; I am hypothesizing in fact that (as a private matter) each of them is strongly committed to protecting the environment. The problem rather has to do with the private character of the decision with which they are faced. Operating within a market economy, they have no choice but to act as private individuals whose acts are independent of the actions of others. If they had a way to decide *together* what they were going to do, they could act in concert to produce the result they all desire; but under a market system all decisions are private ones. Their problem isn't their greed, it's their *isolation*: unable to act together for the goal they all desire, they are forced to act separately and thereby produce a result none of them want.[11] Although Hardin's account is often taken (both by those who accept and those who reject it) as justifying a call to "privatize the commons," this does not seem correct to me. I think he has accurately described a problem endemic to market economies as such, and that its (paradoxical) structure is exactly the structure that Marx described under the name of alienation.

Global warming is a social product: it is the consequence of a series of human decisions and human practices. But it appears to each of the humans who produce it as an unalterable fact independent of their private choices. The problem of the commons arises because my own individual actions, when aggregated with those of my fellow citizens, have public consequences, but since each of us can only act *as* a private individual those consequences appear to each of us as something beyond us, beyond our control and our ability to affect—as an alien power. Global warming is alien in precisely this sense: a process upon which our private choices can have no significant impact and thus to which we can do nothing other than to passively respond—thinking about how to live in a warmer world, worrying about the consequences for future generations, and driving (rationally, and not greedily) each day to work as we do so.

This is the dark side of the "invisible hand": an economy based on private individuals engaging in private transactions with each other in a free market produces a social world whose contours and institutions are the consequences of nobody's choice. Dale Jamieson (chapter 9, this volume), in pointing out the collective action problem raised by global warming, suggests that "conventional morality would have trouble finding anyone to blame" for the harms that climate change may produce, "for no one intended the bad outcome or brought it about," and yet it's worth noticing that this sort of distributed and therefore anonymous responsibility is actually something with which we're quite familiar: it is simply the phenomenon of the market's invisible hand. (Jameson is likely right, though, that conventional morality may not be up to thinking about such a phenomenon clearly.) The difficulty in assigning responsibility for climate change has the same structure as the difficulty explaining the movements of the Dow Jones, or the unemployment rate, or of prices in general: we all are responsible, although for each one of us these things seem like external facts we cannot change but to which we must adjust ourselves.

The environment, I have argued, is "constructed" by our social practices. But under alienation those social practices look to those who engage in them like a series of individual and private ones and so the environment doesn't appear as "our" social product at all. We shape the world, but not in a way we have chosen. Neither the social whole nor the individuals who make it up ever have the opportunity consciously and rationally to decide among alternative visions of what the world could be, or to provide justifications for what their practices bring about. Instead the world takes the form it does as the result of an aggregation of individual acts that occur under the sign of alienation—which is to say, an aggregation of acts each of which paradoxically takes the aggregation itself as given and unalterable. Our own deed, as Marx says, comes to be an alien power over and against us (Marx and Engels 1976, 47).

The problem here is a structural one: it has to do with how our practices are organized, and not with who we are as people or with our moral failings. On this point I differ with some of the other authors in this volume: I do not think we're greedy, not even "modestly" so, unlike Jason Kawall; I do not think our actions are "wanton," unlike Jeremy Bendik-Keymer; and I certainly don't think we're the "scum of the earth," unlike Stephen Gardiner. I think most of us as individuals respond rationally to the social and environmental situation with which we are faced, just as the herdsmen do in Hardin's fable. Bendik-Keymer describes our behavior as marked by "thoughtlessness." But Hardin's herdsmen aren't thoughtless: they understand perfectly well the (tragic) position they are in. And we do too, nowadays—or many of us do. The problem isn't thoughtlessness, it's powerlessness. We're not stupid or selfish: each of us knows that many of our individual actions have harmful environmental consequences when generalized, but we also know that the generalization occurs whether we engage in the actions or not—it's a power over and against us. We have no way to affect or prevent that generalization unless we act *together*, and we have no way to do that (and know that we don't). And so we act, thoughtfully, decently, ungreedily, and thereby help to bring about calamity.[12]

If our alienation is structural, then the solution to it must be structural as well. To overcome it would require finding a way for the social consequences of our actions to themselves be the object of a social, and public, decision and not just the result of a series of private decisions that have merely private, and therefore infinitesimal, consequences. This cannot happen within the market itself, but rather is a matter of politics. In politics, and more precisely in democratic politics, the community makes a decision to act *as* a community and no longer as an aggregate of private individuals. I do not solve the problem of global warming by making a private decision to engage in certain individual practices (refusing to commute, say) but rather by finding a way to decide jointly with others in the community what sorts of social practices *we* (not I) should engage in.

A number of authors in this volume (e.g., Jamieson, Gardiner, Thompson, Kawall, Bendik-Keymer, and others) note the complexity that arises when one tries to affix responsibility for environmental problems that involve collective action dilemmas. As members of a society marked by alienation, I have been suggesting, we are each individually responsible for the environment we inhabit in the sense that it is the result of the sum total of our individual practices and so we have helped to produce it, but *as a society* we have no way to *take* responsibility for the environment by choosing together the sorts of practices we want to engage in. Allen Thompson draws on an idea of Iris Young's to distinguish between "collective" and "shared" responsibility, and that might be helpful here: I bear a portion of the collective responsibility for climate change through my (alienated) actions in a market-driven society, but what's needed is a way for us all to *share* that responsibility through a common politics. Such a common politics oriented toward a taking up of shared responsibility, I would argue, must be democratic, where this is understood as involving a discursive process and not simply the aggregating of individual votes (which would just repeat the alienated structure). In such a process the goal is to develop a communal consensus about "what *we* ought to do," rather than each individual simply thinking by himself or herself about "what I ought to do."

In a free market economy decisions are made by individuals on the basis of their knowledge of market conditions, but those decisions do not appear to them as affecting those conditions: the object of their knowledge is "society," but this appears as something distinct from them. In a democratic discourse, however, individuals know themselves to be part of the very society whose (communal) decisions they are helping to determine. To use Hegelian language, such a discursive process is not merely conscious but *self*-conscious: in it "society" is no longer something the individual confronts as a separate object, but rather a community with which she identifies and whose acts she recognizes as something for which she (too) shares responsibility. A citizen of such a society would be different from an individual in a market society: someone who recognized her own social responsibility, and who thought about what actions would be discursively justifiable to others before acting. And such a society would itself be different as well: a society committed to democracy and no longer marked by alienation.

I am hesitant to say that this is a matter of actors developing new "virtues," because such language suggests the issue here is primarily one of character.[13] I have been arguing that the key issues here are structural, and not characterological. Hardin's herdsmen do not lack virtue, any more than Marx's proletarians or for that matter his capitalists do. What they lack is the opportunity to choose *together* how to act, and as a consequence the acts that they perform, in the aggregate, become an alien power against them—become, to use another Hegelian term, a "second nature" against which they are impotent. If they had such an opportunity, it seems to me, they would relate in a different and better way—not only toward each other but toward "first nature" too, a nature that they would no longer see as separate from them and as something to be either dominated or feared, but rather as something for which they are responsible and which in their every act they inevitably change.

I take it that this is what Jeremy Bendik-Keymer means when he speaks of the importance of "context," and that it is what Paul Hirsch and Bryan Norton have in mind when they say that "thinking like a planet . . . must [be] a collective undertaking" that occurs not so much at the level of the individual as that of "cultural models." But the change that an overcoming of alienation would involve is not best thought of as a change in models or thought or character, because understanding it in that way fails to grasp the centrality of the idea of practice to what I have been arguing. The problem of climate change arises because we have no way to acknowledge the fact that our practices do in fact construct our environment; to overcome that problem would require not changing our minds, but changing those practices. The challenge isn't to get us to become different or more virtuous sorts of people so much as it is to get us to *do different sorts of things*, in a world structured by different sorts of institutions, and in the context of a different sort of economy and a different sort of politics. (See Shockley, chapter 14, this volume.)

I have suggested that overcoming alienation from the environment would require a social order where decisions about the social practices we engage in were made communally and consciously, via democratic discourse, and not left to the nature-like workings of a free market. In such a social order, the environment we would inhabit would be one we would communally *know* to be our communal responsibility. The view of alienation from nature with which I began thinks that the standard for deciding what practices to engage in can be found in "nature," which is to say it wants us to give up our responsibility and allow nature to decide. This does not seem to me to overcome our alienation so much as to perpetuate it: the practices we engage in, and the environment those practices help to produce, remain on such an account subject to something outside of them, and outside of us, just as they are today. The answer is not to abdicate responsibility for our practices and their consequences—we already have done that. It is rather to take back that responsibility from things that are external to us, and in that way end our alienation from an environment that we only mistakenly think is beyond us.

Acknowledgments

An earlier version of this essay appeared as "On Nature and Alienation," in Andrew Biro, ed., *Critical Ecologies: The Frankfurt School and Contemporary Environmental Crises* (Toronto: University of Toronto Press, 2011). I would like to thank Andrew Biro, Jeremy Bendik-Keymer, and Allen Thompson for their helpful comments.

Notes

1. For similar claims, see Bendik-Keymer, chapter 13, this volume; Shockley, chapter 14, this volume; and Hirsch and Norton, chapter 16, this volume.

2. John Stuart Mill's essay "Nature" (1963) remains one of the clearest accounts of these two quite distinct meanings the word "nature" can have, and of the difficulties they each produce. See also Rolston 1986.

3. The whole set of confusions and contradictions here comes to the fore in the philosophical debate about "environmental restoration," particularly (as the first section of this volume makes clear) under conditions of climate change. Both Katz (1997) and Elliott (1997) have serious difficulties explaining what the distinction between the "natural" and the "artificial" really comes to, or why it's important. Hettinger (chapter 1, this volume) and Throop (chapter 2, this volume) both see the difficulty, but it isn't clear to me that they really overcome it. See Vogel 2003.

4. I have altered the translation slightly.

5. It is interesting to consider the rush to geoengineering as a lesser evil (cf. Gardiner, chapter 12, this volume) as itself a result of the kind of alienation Vogel describes, a case where our social decisions force us into alienating relations with our products, here, geoengineering itself.—eds.

6. Hettinger (chapter 1, this volume) suggests that McKibben's claim is too strong, because there are "degrees of naturalness" and one needn't require nature to be "pristine" or "untouched by man." But then to the extent that (nonpristine) nature *has* been touched, it's a human product, and therefore something from which we *can* indeed be alienated in Marx's sense.

7. Should a distinction be drawn here between a product and an effect? We have affected the weather by burning fossil fuels; is it a mistake to say that we have produced it? I don't think so. To recognize that building is never ex nihilo is to see that *things can only produced by affecting other things*. The woodworker affects the wood; the result—the product—is a finished cabinet, but it could also be called wood that shows the effects of her work. Or is the distinction between

product and effect one of intention? But surely one can produce something without intending to do so, as a factory can produce pollution while intending to generate electricity, or as a worker in the factory can help generate electricity while intending to earn a paycheck. "Unintended" products are still products—although often (as I am about to argue) they are products *of alienation*.

8. Some might feel the temptation to seize on the word "helped" here to try to reestablish a dualism, defining nature as "otherness," as that nonhuman factor with which we join or of which we make use in our transformative acts. But I doubt this distinction will do the work defenders of a nature/human dualism want—first because that nonhuman factor is characteristic of *all* human actions, including the construction of nuclear power plants and the burning of fossil fuels, and so no human technologies could possibly be said to threaten or even to ignore or dominate it; and second because the two sides of this dualism remain so intertwined in any action that it makes little sense to try to separate them. Actions don't consist of a human part and a nonhuman part somehow added together (which would be just another way of saying a supernatural part and a natural one): if they did, for instance, what would the "human" part look like *by itself*?

9. Nor does it mean that we have no moral obligations toward it, or that we cannot see ourselves as its "stewards."

10. This is crucial to Marx's distinction between the private, handicraft labor characteristic of precapitalist economies and the sort of labor that takes place in capitalist factories. But the point is really deeper. There is no such thing as a "private" practice, any more than a private language: all practices are what they are because of the norms they follow, and all norms are social norms.

11. Compare Kawall on greed, chapter 11, this volume.

12. Thus Jason Kawall defines (modest) greed as involving excessive pursuit of a good, but does not show that an individual's desire to commute to work by automobile is in itself "excessive." And it would seem difficult to show this, given that the effect on the global climate of such an individual abstaining from commuting would be, essentially, nil.

13. Compare this with the discussion of broadening virtue theory in the introduction to this volume.—eds.

References

Elliott, Robert. 1997. *Faking Nature: The Ethics of Environmental Restoration*. London: Routledge.

Gardiner, Stephen. 2001. The Real Tragedy of the Commons. *Philosophy & Public Affairs* 30 (4): 387–416.

Hardin, Garrett. 1968. The Tragedy of the Commons. *Science* 162: 1243–1248.

Hegel, G. W. F. 1977. *The Phenomenology of Spirit*. New York: Oxford University Press.

Katz, Eric. 1997. The Big Lie. In *Nature as Subject*, 93–107. Lanham, MD: Rowman and Littlefield.

Lukács, Georg. 1971. Reification and the Consciousness of the Proletariat. In *History and Class Consciousness*, 83–222. Cambridge, MA: MIT Press.

Marx, Karl. 1975. Economic and Philosophic Manuscripts of 1844. In *Collected Works*, vol. 3, 229–346. New York: International Publishers.

Marx, Karl. 1977. *Capital*, vol. 1. New York: Vintage Books.

Marx, Karl, and Frederick Engels. 1976. The German Ideology. In *Collected Works*, vol. 5, 19–539. New York: International Publishers.

McKibben, Bill. 1989. *The End of Nature*. New York: Anchor Books.

Mill, John Stuart. 1963. Nature. In *Collected Works*, vol. 10, 373–402. Toronto: University of Toronto Press.

Rolston, Holmes, III. 1986. Can and Ought We to Follow Nature? In *Philosophy Gone Wild*, 30–52. Buffalo, NY: Prometheus Books.

Soper, Kate. 1995. *What Is Nature?* Oxford, UK: Blackwell Publishers.

Vogel, Steven. 2003. The Nature of Artifacts. *Environmental Ethics* 25 (2): 149–168.

16

Thinking like a Planet

Paul D. Hirsch and Bryan G. Norton

Throughout this book, the authors of each chapter have grappled with the question of what constitutes virtue in the context of climatic change and its attendant impacts on ecological and human systems. A central theme has been that our notion of virtue needs to adapt. Thompson and Bendik-Keymer articulate in the book's introduction a *humanist view of adaptation*, which entails "adjusting our conception of who we are to appropriately fit the new global context." Following this line of reasoning in one direction, one might think that adapting virtue is primarily a matter of ethics, and requires rethinking questions of justice, of right and wrong, of what is good and what is evil. In this chapter, however, we will explore the possibility that finding a virtuous path may require adjustment of our *mental models*[1]—our conceptual frameworks and "maps" of the "reality" we act within. We hypothesize that these conceptual frameworks must be adjusted in order to respond virtuously to changing situations. Indeed, we will suggest that, in some cases, the contexts in which individuals develop and act may require not just adopting behavioral patterns that square with moral feelings and a sense of moral appropriateness, but also a recalibration of the understanding of the spatial and temporal context in which action takes place.

Consider, as an example, an attempt to develop an adequate approach to choosing actions and policies for protecting biodiversity at a time when climate is expected to change, species are threatened almost everywhere on earth, and identifying appropriate responses can seem daunting at best. If one assumed that the situation as it obtained in the 1970s when the Endangered Species Act was passed (including the levels of scientific knowledge) still applied, virtuous action might be understood as identifying specific endangered species and developing a "recovery plan" for those species. If expected changes in regional climate patterns unfold as predicted by most climate scientists, however, then responses at the scale of individual species and their habitats will almost certainly fail. A wider spatial context, and a longer sense of time, is necessary to act within the spirit of the Endangered Species Act and protect endangered biodiversity.

This brings us to the thesis of this chapter: key to understanding our obligations and prospects for finding a virtuous path forward in the face of climate change will be a significant shift in our understanding of the context of our moral actions. That shift can best be understood as a metaphor-driven cognitive transformation in the *scale* of the systems we think about and seek to manage. While the relevant scale will vary from problem to problem, for climate change the appropriate scale is surely the planet as a whole. The purpose of the chapter, given the thesis as stated, will be to explore the kind of change that would have to occur in human perspective if we were to undergo such a cognitive transformation and learn to "think like a planet."

We suggest that the way toward a more fruitful perspective from which to seek virtue in a rapidly changing world can be guided by attention to an analogous shift that occurred during the last century. We do this by drawing upon the work of Aldo Leopold, who, in his efforts to manage wildlife in the Southwest Territory of the United States, experienced a shift in his perspective relative to the context in which he acted. In particular, Leopold figured out that the key to understanding the ecological context in which he sought to manage populations of wildlife was to recognize the importance of the spatial and temporal scale at which a given problem is addressed. In his essay "Thinking like a Mountain," Leopold (1949) provided a metaphorical guide to re-modeling a problem at a new scale—the scale of the mountain. In what follows, we will attempt to apply Leopold's lesson to another problem of scale: how can we reconfigure our conceptual models so that, in the face of accelerating climate change, we can understand the expanded scale and context in which we must seek a virtuous path to the future?

Our answer to this question will consist of an argument, in three parts. Part 1 cites an example of an individual transformation, establishing the existence and importance of this kind of cognitive shift and exploring some of the characteristics of such a shift by examining Leopold's transformation in more detail.

Part 2 generalizes the process of individual cognitive transformation to groups and societies, finding that society-wide shifts involve the adoption of new *cultural models*[2] along with the development of new institutions. Examining the case study of the Chesapeake Bay region, we find that it involves a combination of a new metaphor and the emergence of new institutions. This analysis suggests that cognitive transformation at the level of a whole society is more likely in situations in which (1) a new metaphor for understanding a troubled system emerges and calls old models into question; and (2) new institutions arise that respond to, and at the same time create, new models at scales more appropriate to emerging problems.

This leaves us with the question: what *kind* of institutions should these be? And this raises another, equally important question: if they are to move beyond the phase of armchair speculation, how should these institutions *relate* to preexisting structures that modulate political life? In part 3 of the argument we present a rationale and empirical evidence to suggest that institutions supporting our efforts to think like a planet may not usually need to replace existing structures and processes, but rather involve added "layers" organized in polycentric patterns.[3] Beyond these three stages of the argument, we also speculate, in the chapter's conclusion, about what the acceptance of an important role for cognitive transformations means for our understanding of virtue and "virtuous behavior."

Part 1: Cognitive Transformations in Individuals

A forester and wildlife manager in the southwestern United States in the early twentieth century, Aldo Leopold's initial focus was on protecting the economic viability of natural landscapes by ensuring that there were enough deer to attract a steady stream of deer hunters. Knowing that the main predator (other than humans) of deer in the Southwest was the wolf, he formulated a simple plan: exterminate the wolves so there would be abundant deer for hunting. The result was an explosion in the deer population, then a dramatic decline as they overgrazed the available vegetation and succumbed to starvation during hard winters. It was, in the words of Leopold, as if someone had given "God a new pruning shears, and forbidden Him all other exercise. In the end the starved bones of the hoped-for deer herd, dead of its own too-much, bleach with the bones of the dead sage, or molder under the high-lined junipers" (1949, 140).

Were the ecological destruction caused by Leopold and its negative repercussions for the region's economic viability a reflection of Leopold's lack of virtue? Or was it a lack of experience? While this is a point on which philosophers might disagree, the idea we want to focus on is that

Leopold failed to comprehend the spatial and temporal context in which his actions were relevant. Whether his aim of increasing deer availability at the expense of wolves was a virtuous one or not, it is clear that his lack of awareness of the system in which he was acting left him unable to fulfill his intentions, much less protect broader values he was not even considering. After witnessing the deer population explode and the mountainsides lose their vegetation and erode, however, his eyes were opened to the existence of a larger ecological system that included not only predator-prey relationships between hunters, wolves, and deer, but feedback loops between deer populations and mountainside vegetation. This ecological system, in turn, he came to see as embedded in a larger—and slower—geological system that included the process of erosion and the associated loss of topsoil. The transformation in Leopold's thinking is powerfully symbolized by a statement he makes at the outset of the essay that "only the mountain has lived long enough to listen objectively to the howl of a wolf." Only the spatial scale of the mountain is large enough, and the temporal scale long enough, to encompass the ecological and geological dynamics within which any "solution" he might formulate from then on would play out.[4]

Leopold's account provides an important case study in how a conscientious individual, armed with the latest science, can undergo a transformation in the mental models that shape his or her reasoning.[5] We can cite several characteristics of Leopold's transformation that help us to understand what can be expected if and when today's individuals undergo a similar transformation in the mental models that shape thinking about climate change and possible responses to it.

In order to understand the mental model Leopold was using when killing wolves, we can note the similarities between his original model and Gifford Pinchot's[6] understanding of the role of foresters and environmental managers more generally, both of which are based on a metaphor of nature as a productive system. This metaphor is associated with a particular set of goals and strategies for achieving those goals, a particular set of dynamics on which attention is to be focused, and a particular set of values. The primary goal associated with the metaphor of nature as a productive system is one of maximizing productivity in the form of items in demand within the economic system. In thinking through how to meet such a goal, recognizing the limited efficacy of the arid and semi-arid ecosystems of the Southwest for growing timber, Leopold developed the economic development strategy of deer hunting and hunting-related tourism. This goal and way of achieving the goal led Leopold to focus attention on deer populations, and to concern himself with perceived limits to their increasing number as well as threats to their long-term viability. From the model he was working within, he thus concluded that wolves were "vermin." By contrast, the value deer hunters place on high success rates were highlighted in the model.

In his new mental model, created sometime after Leopold's goals were thwarted by the reality of predator-prey relationships and the ecology of mountainside ecosystems, economic productivity is still valued, but its role and functions are embedded in a larger ecological system functioning in the background to support the deer-and-wolf drama as it plays out on the mountain.[7] This expansion in mental models was directly shaped by Leopold's scientific work on deer populations (aided by his colleagues and former colleagues in studying and managing the Southwest's game fields), which provided evidence that his actions to remove wolves and establish large and stable deer populations thwarted his immediate goals and also threatened important values that were excluded from the decision process by the model he assumed. His narrow focus on deer productivity and his means to maximize deer populations blinded him to effects on other values, such as the stability of topsoil and water clarity in streams. Leopold, however, while mentioning the evidence he saw of the impacts of too many deer on vegetation, describes his transformation not as a scientific discovery but as *a shift in the core metaphor* that shaped his model of key dynamics playing out on the mountainsides. As he adopted a new model, the goals, strategies for achieving them, and valuations he placed on the system and elements within it shifted dramatically.

Leopold's experience and changing views are quite relevant to the situation we face today: many people see global climate change in terms of impacts on economic systems. This association of climate change with economic analyses is evident in that evaluations of actions affecting climate change are usually assessed in terms of calculations regarding the impacts of various policies on GDP (see, for example, Nordhaus 1993; Nordhaus 2001; Stern 2008). The analysis provided here suggests that, in learning to think like a planet, we should expect and look for important shifts in the metaphors that shape our view of the systems conceptualized for management.[8] Developing a more appropriate model for understanding climate change must, in other words, involve overcoming the productivity metaphor as applied to the atmosphere and the systems affected by the atmosphere.

Can we, following in Leopold's intellectual footsteps, recognize economic metaphors of productivity as incomplete and too simple, and create metaphors that are more complex, metaphors capable of recalibrating our value systems and readjusting our behaviors and actions as appropriate for the new model we embody in our future thinking? Such a model, or models, would provide a spatial reference system for the evaluation of choices and actions that includes the atmosphere, the oceans, and the poles. Conceptually—just as Leopold's models came to include the relationships among deer populations, vegetation, and eroding topsoil—such models would show the relationships among human activities, greenhouse gases, and patterns of climate. Temporally, they would relate today's activities to dynamics that have unfolded over history and will continue to unfold over the course of decades and generations. In addition, they would differentiate among human activities that contribute in different ways to prospective climatic changes; that may be affected in different ways from their unfolding over time; and that may play different roles in the development of solutions.

Such models, and the metaphors on which they are based, are the essence of what we mean by thinking like a planet. At a very deep level, shifts in such core metaphors are closely related to changes in how we think about virtue. It is their adoption and use—and ongoing development and refinement—that we suggest provides the context for the evaluation of whether our actions can be deemed virtuous or lacking in virtue in our efforts to respond effectively to the challenge of climate change.

Part 2: From Mental Models to Cultural Models and Institutions

Leopold was a remarkable individual. He was unusually curious, and an unmatched observer with far less dogmatism than most people. He also had the benefit of being able to design a policy based on one way of conceptualizing a system, reap the results of his wolf-exterminating "policy experiment," and then revise his conceptualizations of the system's dynamics based on that experience. While there are undoubtedly people alive today (although they may be rare in number) with Leopold's combination of curiosity, powers of observation, and open-mindedness, the issue of climate change is neither understandable, nor experienceable, nor alterable at the individual level.[9] Indeed, to the extent that human activity plays a causal or ameliorating role, or both, significant change can only be the result of the cumulative efforts of large groups of people aggregated over long periods of time.

Thinking like a planet, in order to mean anything, must therefore represent a collective undertaking. More specifically, the expansion of spatiotemporal perspective represented by Leopold's shift must occur not only in individual mental models but also at the level of cultural models, and these new cultural models must be linked in some way with the development of new institutions. Thompson (this volume) has made a similar argument with respect to the evolution of responsibility from the individual to the collective level. In this section, we add to Thompson's discussion by focusing more closely on the cognitive aspects involved in scaling up from individual to collective notions of action and virtue, which will then serve as a basis for our own approach to discussing the need for and obstacles faced by new institutions.

To say that thinking like a planet is best conceived at the level of the collective is not something mysterious, but in fact a way of understanding cognition that cognitive scientists, influenced by parallel research in the philosophy and sociology of science, have been engaging in for several years. The basic reasoning, discussed under the mantle of distributed cognition (Hutchins and Klausen 1996; Giere and Moffatt 2003), and more recently, life-world dependency of cognition (Hoffmann 2007) and environmental perspectives (Nersessian 2006), involves the notion that *external* representations—those occurring outside of the mind—should be considered as part of the thinking process. While we can do simple math in our heads, for example, our ability to engage in mathematical thinking changes the moment we discover that we can use our fingers and toes to count; it changes again when we start to perform manipulations on numbers with paper and pencil; and changes yet again when we start to use calculators and computers (Giere and Moffatt 2003). Each of these activities involves interacting with some form of representation, and whether that representation is in the mind, part of a body, equations on paper or numbers on a screen is less important than its role in facilitating the thinking process. Any discussion of what it means to think, the reasoning goes, is impoverished if it does not include the whole cognitive system of the thinker, his or her body, written symbols, technological instruments, and the relations among them (Hoffmann 2007).

Two examples powerfully illustrate the notion of distributed cognition, and highlight some of its key features in a way that will be useful for subsequent discussion of cultural models and institutions. A canonical example of distributed cognition is the thinking done by an airplane crew in landing a plane (Hutchins and Klausen 1996). Clearly, this is an activity that requires a form of cognition that cannot occur in a single human mind. Only through integration of the information and skills of pilot, crew, and air-traffic controllers—in concert with the information and skills embedded in the plane and its instrumentation—is a successful landing possible. A second example comes from the work of Nersessian and colleagues, who have conducted extensive ethnographical work in interdisciplinary biomedical research laboratories (Nersessian et al. 2003; Nersessian 2006). One finding from their research is that dramatic progress occurs when ideas and objects are developed that are meaningful to thinkers on both sides of a social or intellectual boundary. In the case of the biomedical lab, new diagrams and physical models that combine both organic and engineered elements become a part of the distributed thinking processes that allow biologists and engineers to work together and innovate.

Success in biomedical research labs and passenger airplanes, it is clear, depends on a form of distributed cognition. As complex as planes and laboratories are, the climate system and the multiple social, technical, geological, biological, and other systems with which it interacts are drastically more complex (see Hirsch et al. 2011). If we wish to confront climate change, we must therefore learn to think in a way that is part pilot—acting decisively on what we know; and part interdisciplinary researcher—working across intellectual and cultural boundaries to learn more about what we don't. Just as plane dashboards and hybridized diagrams and models are central to the distributed thinking process of flight crews and interdisciplinary researchers, respectively, central to humanity's ability to adapt to climate change is the adoption of new cultural models—new "maps" of the world and our relationship to it. These new representations allow us to conceive of our collective actions— and their impacts—at scales of space and time that extend beyond those defined by immediate economic concerns.

To serve as the basis for learning and action in political contexts, new cultural models must be closely associated with the development of new *institutions*, in particular institutions that function to adjust the boundaries of the system to be managed. In their most basic sense, institutions are "the external (to the mind) mechanisms individuals create to structure and order the environment" (North and Denzau 1994). Through institutions, our ideas about how the world works, and what is necessary to act within it, are articulated in language, instantiated into rules and structures, and to a greater or lesser extent empowered (or resisted) by the instruments of the state, business, or civil society. Institutions are essential to create a "public," in John Dewey's sense (1927): an organic society capable of experimenting, observing and learning in the face of threats and problems.[10]

The emergence of shared cultural models and associated institutions—both reflective of a deeper cognitive shift—can be illustrated in the case of the development of a new "map" of the Chesapeake Bay watershed and the creation of new institutions based on this new map. The shift in understanding that occurred in relation to the pollution of the Chesapeake is analogous to Leopold's shift, with the difference being that it occurred at the societal level and was closely linked to the development of lasting institutions. The Chesapeake Bay—the largest estuary in the United States and one of the United States' most productive aquatic ecosystems—has been in decline since the late 1960s. Originally, the main threat was thought to be activities occurring within or nearby the Bay itself (oil spills, sewage outflows, and toxic pollutants from nearby industries). Starting in the 1970s, however, a steady stream of scientific research determined that the primary threat was in fact excess nutrients from agricultural, urban, and sewage runoff spread out over the states of New York, Pennsylvania, Delaware, West Virginia, Virginia, and Maryland, and Washington, DC.

The science that emerged in the late 1970s and early 1980s regarding the Bay's close connection to the tributaries that bring it water from far-away fields and roads led to a shift in the cultural models of people living in those states, and an associated expansion in the spatial scale considered relevant for preserving the integrity of the Bay (EPA 1983; Horton 1987). This conceptual shift was closely linked to the development of a new set of institutions, including the multistate Chesapeake Bay Agreement, initiated in 1983 and revised in 1987 and 2000 and, equally important, the emergence of the Chesapeake Bay Foundation as a highly effective nongovernmental organization. The agreement constitutes a new way of managing the problem by protecting the Bay in the context of its 64,000-square-mile watershed that covers parts of five states.[11] Interestingly, following a flagship study initiated in 1976, the EPA published a booklet intended to explain its findings and their implications for managing the Bay, called "Chesapeake Bay: Introduction to an Ecosystem" (EPA 1982). The preface to this document, signed by both the director and deputy director of EPA's Chesapeake Bay Program, says: "The Bay is, in many ways, like an incredibly complex living organism. Each of its parts is related to its other parts in a web of dependencies and support systems. For us to manage the Bay's resources well, we must first understand how it functions." The transformation that took place across the Chesapeake region, then, is similar to Leopold's transformation in that it involved a radical redrawing of the boundaries of a managed ecosystem, and that redrawing was explained and presented as a change, not in some specific belief, but rather in the adoption of a new basic metaphor for understanding the system.

The development of the Chesapeake Bay Agreement presents a model, perhaps, for global-level institutions designed to respond to the threat of climate change. The Kyoto Protocol, and the October 2011 Cancun Climate Conference (and whatever emerges from it) represent early attempts at linking institutions to the growing awareness that addressing climate change means working across national boundaries. While the development of necessary institutions is off to a slow start, we nevertheless believe that the emergence of a new and more appropriate cultural model for living in a world threatened by catastrophic climate change depends upon the development of effective institutions that are appropriate to the expanded metaphor of thinking like a planet.[12] The role of institutions, we hypothesize, is important because the cultural models operative in most societies for thinking at the global scale are inchoate but—given the drumbeat of demand for economic growth and jobs—ultimately shaped by default to the economic metaphor of earth as a productive system. If we are to learn to think like a planet, the new metaphors that are adopted will surely affect individuals and their behavior. Yet considering the scale of the problems involved, it seems certain that the necessary shift in cultural models will be associated with developing new and more appropriately scaled institutions and organizations.

This proposal is presented with a caveat, however. It is by no means our intention to suggest that thinking like a planet entails some form of global governance and associated dissolution of the nation-state. This would likely be seriously problematic both from the perspective of political expediency and from the perspective of promoting human virtue. In the next section, we address the question of the appropriate relationship between emerging institutions designed to facilitate thinking like a planet and the preexisting institutions—such as states and countries—that have evolved over centuries, are central to people's identities, and embody humanity's best efforts to date to ensure virtuous ideals such as freedom, security, and democracy.

Part 3: Expanding Boundaries and Political Realities

A central issue that makes thinking like a planet both necessary and difficult is the fact that the boundaries of environmental problems transgress preexisting socially and politically defined boundaries—boundaries
defined by neighborhoods, communities, counties, states, nations, and
the variety of organizations that operate within and among them. The
result of this mismatch is that there is an institutional void in society's
ability to deal effectively with problems that emerge in systems larger
than available jurisdictions (Levin 2000; Hajer 2003; Folke et al. 2007).
This was clearly evident in the case of the Chesapeake before the signing
of the Chesapeake Bay agreement and the emergence of the Chesapeake
Bay Foundation.

But this should not be taken to mean that there is a "right" scale at
which to confront a problem. Indeed, more recent research suggests that
the airshed of the Chesapeake Bay, which includes Midwestern states
from which factory emissions make their way into the Bay via patterns
of precipitation, is the appropriate scale at which to manage the Chesapeake. The question of what is the "right" scale to manage the Chesapeake is furthermore complicated by the fact that different actors may
identify with different scales, and may also benefit or be harmed when
a problem, particularly one around which lasting institutions are built,
is defined at a particular scale (Williams 1999; Kurtz 2003; Lebel,
Garden, and Imamura 2005). It is important, therefore, to recognize the
"wicked" nature of such problems (Rittel and Webber 1973; Norton
2005)—the key aspect of wickedness being that people and groups with
different values and perspectives articulate the nature of the problem in
different ways.[13]

Today, climate change presents perhaps the most dramatic example of
a mismatch in scale between the boundaries defined by the physical
dynamics of a problem and existing socially and politically defined
boundaries. Climate change is also a wicked problem in that people living
in the global North and South understand the problem quite differently,
as do people with different levels of trust in science, attitudes toward
government, and religious faith. It is no wonder, then, that the push to
develop new institutions which function at the scale of climate change—
especially those which purport to exercise some form of control or
authority—has led to pushback. In certain cases, this pushback can be
associated with denial of the science underlying problems like climate
change, and in extreme cases a rejection of science as a meaningful guide

for human thinking and activity. When this happens—particularly if, as we claim, virtue depends at least in part on understanding the context in which actions are undertaken (compare Kawall, chapter 11, this volume)—it seems increasingly unlikely that virtuous behavior will emerge.

Nevertheless, it is important to point out that such pushback may represent people's striving to fulfill widely held and longstanding notions of virtue and responsibility. For example, many individual citizens, social groups, and policymakers are reluctant to cede decision making or political authority to emerging institutions they may perceive as inhibiting sovereignty, freedom, or creativity. If we accept the legitimacy of multiple values in shaping the development of institutions, therefore, it is important to distinguish between people's ability to *conceive* of the spatial and temporal scale in which their actions are meaningful, and their willingness to *cede authority* to institutions designed to function at the scale.

A brief empirical example demonstrates this distinction: as in the Chesapeake Bay, water problems in the state of Georgia transgress political jurisdictions. Before 2005, the bulk of water planning was done at the level of individual counties, and it was thus difficult to account for the ways in which decisions made by an upstream county affected the opportunities and quality of life downstream—decision makers failed to "think like a watershed." In 2005, a legislatively mandated advisory process commenced in which local government officials, water managers, representatives of farming and environmental groups, and citizens participated in the chance to redraw Georgia's water planning boundaries. In essence stakeholders were given the opportunity to use watersheds as a metaphor in the development of new cultural models and associated institutions to structure water planning in the state.

In a survey we conducted of participants to the advisory process, although a strong majority (71 percent) favored the development of water management institutions that transcend the borders of counties and bring people together at the watershed scale, the percentage willing to assign political authority to those regions was significantly lower (51 percent) (Hirsch 2008). This seeming split between those willing—and those unwilling—to assign political authority to watershed-based institutions resulted in a policy outcome that is not uncommon in efforts to restructure state and local planning according to the contours of an environmental problem. In the final maps that came out of the Georgia process, watershed boundaries were partially followed as a means to structure planning, but—wherever they conflicted with the boundaries of counties or metropolitan Atlanta (the economic driver of the state)—watershed boundaries were deemphasized and boundaries demarcating counties and economically important regions came to the fore (see Hirsch 2008). One can not help but think that, if there had been a clearer distinction made between developing collaborative institutions for thinking like a watershed on the one hand, and devolving the organs of the state to a watershed-based regulatory authority on the other, then a "map" might have emerged from the process that more fully allows water professionals and county decision makers to account for the upstream and downstream contexts of their choices. Such a map would then function like the dashboard of a plane or a hybrid laboratory model, enhancing the ability of stakeholders and decision makers to think together at a larger and more appropriate scale.

This example demonstrates the difference between *thinking* at a certain scale and ceding decision-making power to larger-scaled political bodies. While the clear majority of those who participated in the process understood the physical dynamics of watersheds and were willing to participate in institutions that were structured by them, the results of the survey indicate the importance to Georgia's citizens of maintaining the authority of states and counties. From the standpoint of both political expediency and the importance of the virtues we look to preexisting jurisdictions to uphold, the implication is that new institutional structures should perhaps not be conceived as replacing old ones; rather, they might be conceived of as "layers" that constructively complement existing entities. This idea is supported by Ostrom and colleagues, who articulate a role for "polycentricity" in natural resource management (Andersson and Ostrom 2008; Ostrom 2009). In essence, polycentrism refers to overlapping jurisdictions, or the organization of a region or a set of issues around multiple "centers." While some of these centers may be imbued with the power to regulate and enforce law, others might play a role primarily in information sharing or the convening of deliberations (Guston 2001). These kinds of institutions need not threaten the functioning of governance as it plays out within the divisions of county, state, and nation.

Perhaps unlike any previous environmental issue, the institutional arrangements that may emerge from current climate change initiatives have the potential to allow multiple people across multiple nations and sectors to think like a planet about problems that manifest at the scale of the planet. At the same time, they have the potential to be rendered inconsequential—or even cause more harm than good—if they are perceived as threatening the right and ability of people to make decisions that are meaningful and virtuous within their particular contexts, according to values they may have fought and died—or at least voted—for. We do not believe it is a stretch to say that some of the resistance to the science of climate change witnessed over the last decade is motivated by a grave concern for its political implications. Whether that is expressed as condemnation of communism or fascism or authoritarian global government run by technocrats, in each case there is resistance to an ideology that is seen as favoring collective goods over individual freedoms.

It may be, therefore, that institutions that support thinking like a planet may be most viable when conceived of as complementary to— rather than a replacement for—existing structures that organize decision making and social life. That is, while the *thinking* necessary to confront environmental problems such as climate change can be greatly augmented by insights and lessons from ecology, geology, and other sciences, much of the social machinery for deciding, regulating, and enforcing decisions may be best off where it currently resides.

In the concluding section, it will be important to return to a discussion of how thinking like a planet, so long as it is not usurped by overly politicized interpretations, can inform and support the development and expression of human virtue.

Conclusion: Virtue in a Rapidly Changing World

Articulating a role for mental and cultural models raises interesting and important questions about how we think about virtue and our judgments regarding whether individuals are acting virtuously in changing situations, including our current struggle to chart a moral course of action in the face of accelerating climatic and ecological change. Key among these questions is whether using an inappropriate model of the dynamics under consideration is a matter requiring more virtue or, alternatively, if it is possible to respect and recognize actions as virtuous within the context in which they are taken and given the current state of knowledge. Grasping the latter option, we have argued that developing virtue in an age of climate change involves not only ethical but also cognitive aspects.

Looking back to Leopold's decision to exterminate wolves, it is not difficult to respect his conscientious pursuit of economic development through maximizing deer populations for hunters. Leopold, working within the purview of the U.S. Forest Service's utilitarian management framework as articulated by Pinchot, would have seemed at the time to be a highly virtuous actor in the social endeavor of managing resources responsibly. Indeed, Leopold was, by all accounts, a conscientious, intelligent, and effective actor in the situations he found himself in. He was also a "good citizen," and participated in local economic development organizations.

With the advantage of hindsight, we can say that Leopold's actions turned out to be shortsighted and, especially, unduly constrained by a narrow focus on economic values associated with a deeply engrained metaphor of nature as a productive system. Leopold, however, in an almost-heroic act of scientific reasoning and creative speculation, found his favored actions to be inappropriate when judged by their results and within a new perspective. His thinking and perspective—including the metaphors he used, the mental model he constructed, and the goals of his overall managerial efforts—underwent a cognitive transformation, and as a result he adopted new goals and engaged in new activities. In 1935, for example, he cofounded The Wilderness Society, which included as part of its platform predator protection in wilderness areas.

Leopold's transformation and associated behavior change can be understood as an instance of the adaptation of virtue, and provides grist for the mill for further inquiry and even empirical work on the relative roles of ethical and cognitive processes, and the relationship between them, necessary for developing a humanist view of adaptation. We conclude by posing one additional speculation as to the nature of this relationship by pointing out that, in making the shift to thinking like a mountain, Leopold did not lose sight of the importance of the individual and the particular. On the contrary, in light of his new understanding, wolves—previously conceptualized as vermin—take on a value that they did not have before, a value that involves something more than just functionality. The narrative in which he communicates his expanded understanding at the scale of the mountain is explicitly linked to an appreciation, verging on compassion, for the "green fire" dying in the eyes of a she-wolf he had just gunned down:

We reached the old wolf in time to watch a fierce green fire dying in her eyes. I realized then, and have known ever since, that there was something new to me in those eyes—something known only to her and to the mountain. I was young then, and full of trigger-itch; I thought that because fewer wolves meant more deer, that no wolves would mean hunters' paradise. But after seeing the green fire die, I sensed that neither the wolf nor the mountain agreed with such a view. (Leopold 1949, 138)

It seems fair to say then that even as Leopold's sense of context expanded to the scale of the mountain, his sense of understanding of and empathy for the various elements that comprise the mountain became more attuned and refined. Starting from this recognition, it becomes clear that cultural models and associated institutions for thinking like a planet may have very different things to offer than the specter of global government: for starters they can set the stage for fuller expression of the virtues people already hold. Whether, as Sandler (chapter 3) and other authors of this volume suggest, new kinds of virtues such as openness and reconciliation are also required we leave as an open question for further researchers and scholars, not to mention global citizens.

Acknowledgments

The empirical research presented in part 3 of this chapter was supported by the Human Social Dynamics Program of the National Science Foundation (NSF Award 0433165).

Notes

1. A "mental model" can be understood as an internal representation of how something works in the world; how different dynamics are interrelated; and how a person's actions and experience shape and are shaped by those dynamics (see Doyle and Ford 1998 and Johnson-Laird 2000).

2. Cultural models are mental models that are shared by groups, making communication about shared values and goals possible (see Kempton, Boster, and Hartley 1996).

3. A conclusion that seems at odds with Shockley's position (chapter 14, this volume).—eds.

4. It is not a stretch to replace the "mountain" metaphor with the idea of an "ecosystem," in which case Leopold's famous story can be seen as the dawn of the idea of ecosystem management in American conservation.

5. Does the cognitive malleability in stories such as this decrease Bendik-Keymer's worry (chapter 13, this volume) that our cognitive nature may be tragically limited in the face of large-scale ecological problems?—eds.

6. Gifford Pinchot, the first chief of the U.S. Forest Service, was a manager generally associated with the notion that managing natural resources means maximizing their economically efficient use.

7. Compare what Schlosberg (chapter 8, this volume) urges.—eds.

8. Compare this with the debate about "historicity" ranging across part I of this volume and Higgs, chapter 4, especially.—eds.

9. Higgs (chapter 4, this volume) claims we may not have the "ecological imagination" to anticipate it.—eds.

10. One question is whether institutions at the same time involve heteronomy for their members, as Shockley (chapter 14, this volume) claims.—eds.

11. This is *not* to say that all of the environmental problems and political disputes have been happily solved in the Chesapeake region. Our claim is only that, in spite of continuing problems, the public, environmentalists, and decision makers are now in a better position to understand and address the problems than they were when they shared a simpler and smaller-scale mental model of the Bay ecosystem.

12. What would be the role of ecological history in such institutions? Compare with part I of this volume.—eds.

13. Vogel (chapter 15, this volume) might also note the *alienated* nature of these problems—whenever social realities are assumed as natural, for example, as in the case of the nation-state.—eds.

References

Andersson, K., and E. Ostrom. 2008. Analyzing Decentralized Resource Regimes from a Polycentric Perspective. *Policy Sciences* 41 (1): 71–93.

Dewey, J. 1927. *The Public and Its Problems.* Athens, OH: Swallow Press.

Doyle, J., and D. Ford. 1998. Mental Models Concepts for System Dynamics Research. *System Dynamics Review* 14 (1): 3–29.

EPA. 1982. "Chesapeake Bay: Introduction to an Ecosystem." U.S. Environmental Protection Agency, Chesapeake Bay Program, Philadelphia.

EPA. 1983. "Chesapeake Bay: A Framework for Action." U.S. Environmental Protection Agency, Chesapeake Bay Program, Philadelphia.

Folke, C., L. Pritchard, F. Berkes, J. Colding, and U. Svedin. 2007. The Problem of Fit between Ecosystems and Institutions: Ten Years Later. *Ecology and Society* 12 (1): 30.

Giere, R., and B. Moffatt. 2003. Distributed Cognition: Where the Cognitive and the Social Merge. *Social Studies of Science* 33 (2): 301.

Guston, D. 2001. Boundary Organizations in Environmental Policy and Science: An Introduction. *Science, Technology & Human Values* 26 (4): 399–408.

Hajer, M. 2003. Policy without Polity? Policy Analysis and the Institutional Void. *Policy Sciences* 36 (2): 175–195.

Hirsch, P. D. 2008. "Making Space for Environmental Problem Solving: A Study of the Role of 'Place' in Boundary Choices Using Georgia's Statewide Planning Process as a Case." Ph.D. Dissertation. Georgia Institute of Technology.

Hirsch, P. D., W. A. Adams, J. P. Brosius, A. Zia, N. Bariola, and J. L. Dammert. 2011. Acknowledging Conservation Trade-offs and Embracing Complexity. *Conservation Biology* 25: 259–264.

Hoffmann, M. 2007. Learning from People, Things, and Signs. *Studies in Philosophy and Education* 26 (3): 185–204.

Horton, T. 1987. Remapping the Chesapeake. *New American Land* 7 (4): 7–26.

Hutchins, E., and T. Klausen. 1996. Distributed Cognition in an Airline Cockpit. In *Cognition and Communication at Work*, ed. Y. Engestrom and D. Middleton, 15–34. Cambridge, UK: Cambridge University Press.

Johnson-Laird, P. 2000. The Current State of the Mental Model Theory. In *Mental Models in Reasoning*, ed. M. Labra, N. Lopez, and J. Garcia Madruga, 17–40. Madrid, Spain: Universidad Nacional de Educación a Distancia.

Kempton, W., J. Boster, and J. Hartley. 1996. *Environmental Values in American Culture*. Cambridge, MA: The MIT Press.

Kurtz, H. 2003. Scale Frames and Counter-scale Frames: Constructing the Problem of Environmental Injustice. *Political Geography* 22 (8): 887–916.

Lebel, L., P. Garden, and M. Imamura. 2005. The Politics of Scale, Position, and Place in the Governance of Water Resources in the Mekong Region. *Ecology and Society* 10 (2): 18.

Leopold, A. 1949. Thinking like a Mountain. In *A Sand County Almanac*, 129–133. New York: Ballantine Books.

Levin, S. 2000. Multiple Scales and the Maintenance of Biodiversity. *Ecosystems* 3 (6): 498–506.

Nersessian, N. 2006. The Cognitive-Cultural Systems of the Research Laboratory. *Organization Studies* 27 (1): 125–145.

Nersessian, N., E. Kurz-Milcke, W. Newstetter, and J. Davies. 2003. Research Laboratories as Evolving Distributed Cognitive Systems. In *Proceedings of the Twenty-Fifth Annual Conference of the Cognitive Science Society*, ed. R. Aterman and D. Kirsh, 857–862. Hillsdale, NJ: Erlbaum.

Nordhaus, W. 1993. Reflections on the Economics of Climate Change. *Journal of Economic Perspectives* 7 (4): 11–25.

Nordhaus, W. 2001. Climate Change: Global Warming Economics. *Science* 294 (5545): 1283.

North, D., and A. Denzau. 1994. Shared Mental Models: Ideologies and Institutions. *Kyklos* 47 (1): 3–31.

Norton, B. 2005. *Sustainability: A Philosophy of Adaptive Ecosystem Management*. Chicago: University of Chicago Press.

Ostrom, E. 2009. "A Polycentric Approach for Coping with Climate Change." World Bank Policy Research Working Paper Series No. 5095 (October 1, 2009). Background paper for the WDR 2010.

Rittel, H., and M. Webber. 1973. Dilemmas in a General Theory of Planning. *Policy Sciences* 4 (2): 155–169.

Stern, N. 2008. The Economics of Climate Change. *American Economic Review* 98 (2): 1–37.

Williams, R. 1999. Environmental Injustice in America and Its Politics of Scale. *Political Geography* 18 (1): 49–73.

About the Contributors

Jeremy Bendik-Keymer is Beamer-Schneider Professor in Ethics and Associate Professor of Philosophy at Case Western Reserve University

Stephen M. Gardiner is Professor of Philosophy and Ben Rabinowitz Professor of the Human Dimensions of the Environment at the University of Washington, Seattle

Ned Hettinger is Professor in the Department of Philosophy at the College of Charleston

Eric Higgs is Professor in the School of Environmental Studies at the University of Victoria

Paul D. Hirsch is Assistant Professor in the Department of Environmental Studies at the State University of New York College of Environmental Science and Forestry

Breena Holland is Associate Professor in the Department of Political Science and the Environmental Initiative at Lehigh University

Dale Jamieson is Professor of Environmental Studies and Philosophy at New York University

Jason Kawall is Associate Professor in the Department of Philosophy and in Environmental Studies at Colgate University

Jozef Keulartz is Associate Professor in the Department of Applied Philosophy at Wageningen University and Special Professor for Environmental Philosophy at Radboud University

Andrew Light is Associate Director of the Institute for Philosophy and Public Policy at George Mason University and Senior Fellow and Director of Climate Policy at the Center for American Progress

Bryan G. Norton is Distinguished Professor in the School of Public Policy at Georgia Institute of Technology

Ronald Sandler is Associate Professor of Philosophy and Director of the Ethics Institute at Northeastern University

David Schlosberg is Professor in the Department of Government and International Relations at the University of Sydney

Kenneth Shockley   is Associate Professor in the Department of Philosophy at the University at Buffalo, State University of New York

Jac. A. A. Swart   is Associate Professor with the Faculty of Mathematics and Natural Sciences at the University of Groningen

Allen Thompson   is Assistant Professor in the Department of Philosophy and in the Environmental Humanities Initiative at Oregon State University

William M. Throop   is Professor of Environmental Studies and Philosophy at Green Mountain College

Steven Vogel   is Brickman-Shannon Professor in the Department of Philosophy at Denison University

Index